# ICE AGE MYSTERY

## A PROPOSED THEORY
## FOR CLIMATE CHANGE

by

LG BELL, P Eng

Bloomington, IN        Milton Keynes, UK

authorHOUSE®

AuthorHouse™
1663 Liberty Drive, Suite 200
Bloomington, IN 47403
www.authorhouse.com
Phone: 1-800-839-8640

AuthorHouse™ UK Ltd.
500 Avebury Boulevard
Central Milton Keynes, MK9 2BE
www.authorhouse.co.uk
Phone: 08001974150

First published by AuthorHouse 8/21/2007

ISBN: 978-1-4259-9923-0 (sc)

Printed in the United States of America
Bloomington, Indiana

This book is printed on acid-free paper.

# ACKNOWLEDGMENTS

I would like to thank the following people who assisted, one way or another and perhaps inadvertently, in the development of the ideas for this book, or in the decision to publish.

John C. Anderson, David Barber, Andre Berger, Roy Bishop, George Garland, Bill Gough, Al Gore, Hartmut Grassl, Henry Hengeveld, Kaz Higuchi, Roy Koerner, Roberta McCarthy, Ted Munn, Guy Nason, Manish Patel, Paul Renne, Archie Robertson, Tony Sawatzky, Ted Shepherd, Dmitry Sonechkin, Pietre Tans, Morley Thomas, and Jan Veizer.

There are also legions of nameless to well known individuals who collected data in uncomfortable circumstances that have to be recognized for their contribution. Without their data the story could not have been told.

I give special thanks to Theoretical and Applied Climatology for permission to republish verbatim the first three reports in the series included in Chapters 1,2 and 3 herein. I also thank various journals, referenced at the end of reports, for figures copied from their publications.

# INTRODUCTION

In October 1994 my wife and I attended an Elderhostel event at Killarney Lodge on the northeast shore of Georgian Bay. We had a good time. One of the lecture sessions was on painting with watercolors and a second was on geology with particular attention paid to local rock formations and evidence of glaciation. We were also taken sailing on the bay and canoeing on Lake St. George. It was all very enjoyable and I was particularly interested in the evidence of glaciation. Along with other adventures, the lecturer George Garland, a retired University of Toronto professor, took us on a field trip through the woods not far from the lodge and showed us a moulin. That is the French word for mill and anyone seeing such an artifact would immediately see why it would have been given this name.

A granite outcrop is to be observed on a slope about 10 meters above the valley floor. On the left side there is a groove in the outcrop running up the hill at about 30 degrees to the horizontal that looks like it might be part of a cylinder about 4 meters in diameter. Professor Garland explained that toward the end of the Ice Age there was apparently a pipe running through the glacier carrying water under high pressure. The pipe was maintained by a balance between erosion of the ice by the water and stones flowing through it, on one hand, and deformation of the ice by the pressure of perhaps a few kilometers of ice above, on the other. By chance the pipe wandered to the outcrop and wore the groove in the solid rock.

Slightly to the right of the groove and below the outcrop a hole about a meter in diameter had been drilled vertically into solid granite. The hole is cylindrical, slightly barrel-shaped toward the top, and about three meters deep. There is rubble in the bottom so it is hard to say how deep the hole is. Apparently an eddy was set up by interference of the flow in the pipe by the outcrop, and rocks that happened to be caught in the eddy were induced to drill the well. As the rocks that were trapped were worn away more were added. It is said that a spherical pestle rock can usually be found in the bottom of one of these wells, but we couldn't see if it was there because of the rubble.

I assumed at the time that it would have taken hundreds or even thousands of years of milling to produce such a spectacular artifact. I did eventually, however, find a reference in a book on glaciation that claimed moulins were formed in a "relatively short time". It seems necessary to conclude however that the glacier could not have been moving, because if it moved surely the drilling would be discontinued. I had to leave it at that but the experience started me thinking about other glaciation artifacts and indeed about how the Ice Age could have come to an end.

The Niagara escarpment is a feature of the topography of southern Ontario. It runs more or less up the middle of the southern part of the province with the river flowing over it at Niagara Falls. From there it winds along the Niagara Peninsula to Hamilton then north to a point southwest of Collingwood then west and north up the Bruce Peninsula. A layer of limestone on the west, slightly more resistant to erosion, overbears softer layers that are more easily eroded. This is what produces the escarpment a few hundred meters high. The underlying rock from the shores of Georgian Bay to Lake Ontario, both above and below the escarpment, is limestone.

Since the escarpment would have affected the movement of the ice sheet, it not surprising that there is a difference in glaciation artifacts. The drive south from Collingwood along Airport Road, east of the escarpment, is like being on a roller coaster. In the ups and downs it is to be noted that the soil is rich with no erratics to be seen along the highway. It is not at all like a typical landscape, where the topography has been determined by the erosion of streams and rivers, the soil seems to have been just ploughed up and dumped there.

When you travel north from Brampton on highway #10, somewhat to the west of Airport Road, things are quite different, the first major topographical feature is the Caledon hill that is being excavated by a gravel company. The underlying rock for hundreds of kilometers around is limestone, unsuitable for gravel, so granite rubble must have been brought in. With Thornbury, on Georgian Bay west of Collngwood, as your objective, follow #10 and find your way to highway #124, turn left on highway

#4, then right on #2. Most of this time you are on top of the escarpment. There are again no volcanic erratics. There are a few weather beaten lumps of limestone that appear not to have been put there by man and thus to be erratics. Presumably they were ripped off the upper layers of the escarpment by the glacier and left scattered on the plain. Being soft limestone they erode much faster than granite. On #2 highway one eventually comes to a place called Kolapore, apparently an abandoned town with only a church left to mark the spot. A kilometer or so north of the church the Bruce Trail crosses the highway. This walking trail runs from Niagara Falls to the northern tip of the Bruce Peninsula along the escarpment wherever possible and crosses #2 from east to west north of Kolapore. The trail to the west of the junction follows for some distance along the top of a moraine, which is made up of boulders, mostly limestone but with a few smaller granite rocks in the mix. It is relatively small for a moraine, being only about 40 meters high and 100 meters wide. Travel north again on #2 is mostly downhill to the town of Thornbury and the grand vista of Beaver Valley is on the left. There is no definite precipice in the landscape as is usual along the escarpment, the elevation decreases in stages down to Georgian Bay.

About 20 kilometers west from Thornbury on #26, the escarpment looms to the south fairly close to the highway, eventually a turnoff to the right leads down about a kilometer to Christy Beach on Georgian Bay. From the beach and more spectacularly from cottage properties along Sunset Boulevard to the east, one observes many thousands of granite or other volcanic erratics. They are in all shades from black to gray, brown to orange, some mottled, some not, but all well rounded and with diameters up to as much as 2 meters. Figure 0-1 shows a view from one of the cottages. The rocks have been collected into spits at extensions of property boundaries to provide access to the lake. (As an interesting aside the erosion of these rocks and the action of the prevailing northwest wind of the Bay is probably what provided the sands of the Wasaga beaches).

*Figure 0-1. View of Georgian Bay from one of the cottages on Sunset Boulevard showing spit of granite and other volcanic erratics.*

Travelling further east along the shore road it is to be observed that the land is being cleared for a golf course. Perhaps thousands or even millions of tonnes of boulders have been cleared and hauled away over the last few years to make way for fairways. As one again approaches Thornbury, in the middle of the Beaver River valley, there are no erratics along the shore and the escarpment is out of sight far to the south. The erosion of two river systems apparently caused the fragile point of the escarpment southeast of Thornbury. This provided a ramp that allowed smaller granite erratics to make it to the top and move on to Kolapore to produce the moraine that is found there. Further to the east, there is a trailer park along the shore. We camped there when the kids were small and found the shore to be solid limestone with no boulders. It has been noticed recently that the escarpment, with its famous Blue Mountain ski runs, is very close by on the south side of the highway. Presumably at this point boulders were deposited out in deep water. Just before one comes to Collingwood, there is a spot where erratics can again be seen along the shore and if one should look to the south at this point one can again observe the escarpment in the far distance.

In addition to the moulin mentioned above, there are many artifacts of the melting of the ice sheet in southern Ontario. There is, for example, an esker running north and south through the town of Ajax and spectacular potholes on the Eramosa River, 20 kilometers or so east of Guelph. All of these artifacts are evidence of melting when the ice sheet was stationary. There is also the Rough Moraine, that runs from the Caledon hill eastward across the top of Metropolitan Toronto.

Glaciers do flow downhill in mountainous areas, and moraines are found where they melt. In Ontario the land is more or less flat. Some flow no doubt would result from the sheer weight of the ice buildup, hence the Rouge Moraine, but it is hard to imagine that it would flow up and over large obstacles. What apparently happened is as follows: The climate became warmer as the Ice Age was ending, the ice sheet became more plastic and started to melt but it did not melt uniformly. Perhaps within 500 miles or so of the southern edge winter conditions prevailed because moist air coming from the south would have been cooled by rising to clear the ice. Snow continued to fall in this area and to thus maintain the ice thickness. Farther north however the ice did melt because there was less cloud and less snowfall. Eventually there was a big lake a kilometer or so deep, melted by the effects of waves and buoyancy right down to the ground. This lake is well known to science and is called Lake Agassiz. Eventually melting by the Sun on the southern face of the dam, the ice becoming more plastic and the melting, undermining, and buoyancy effect by water on the northern face, allowed the water pressure to move the dam.

When it started to move, the dam was maybe two kilometers thick and wider, north to south, than the length of Georgian Bay. As it came through the mountains to the north many large chunks of igneous rock were dislodged and carried away. Grinding along together they were made round and smooth. As the glacier bulldozed its way across Georgian Bay, now no longer frozen solid, it picked up the sediment from the bottom, which became mixed with ice and effectively part of the dam. The boulders being heavy settled for the most part into the bottom layers.

Ice under pressure is plastic and when the moving dam encountered the escarpment the lower layers were halted leaving a pile of boulders, not necessarily at the obstruction but close to it, while upper layers above the escarpment were pushed farther south. Around the towns of Thornbury and Meaford you may see the occasional granite erratic but not piles along the shore. This is because the towns are situated in valleys where the escarpment has been pushed farther south by erosion. At these locations the boulders rolled on and were carried farther up the valleys before being deposited. Pressure in the ice from the close proximity of Blue Mountain is presumably what prevented an accumulation at the trailer park. Collingwood is slightly to the east of the escarpment and again there are no erratics along the shore. At this location, without an escarpment to encounter, the boulders ground on and were covered by Georgian Bay silt. It was presumably this silt that created the unusual topography along Airport Road.

The dam did not break and cause a flood as it did out west between the Coast and Rocky Mountain ranges. It started to leak leaving high-pressure water effects such as eskers, moulins, and potholes. In spite of pressure from the weight of ice, water creeps, centimeter by centimeter, along the muddy bottom of the dam until it makes its way through. Once there is a flow the channel can migrate into a pipe away from the land as it did at Killarnie, to a river bed to produce potholes in the limestone bedrock as on the Eramosa, or to a self-made riverbed, an esker, above the plain as at Ajax.

This may explain the evidence of the demise of the glacier but not why the climate got warmer to start it melting. In conventional climatology periodic glaciation is attributed to changes in the energy from the Sun, the result of interaction of the orbits of various members of the solar system. Even though periods of glaciation were reputed to be occurring at fairly regular intervals of 100,000yr, about the same as the Earth orbit eccentricity cycle, this theory did not seem credible. For one thing energy from the Sun is only changed by a few percentage points by any of the known Earth-attitude cycles, even though they might occasionally work in concert. In addition, for most of the known history back to 550 million years, there were no glaciation cycles. The attitude cycles have always been there. So, if the influence of external bodies is so powerful, why wouldn't the climate always have cycled?

To solve the problem I decided to concentrate on the mystery of the Ice Age ending. I thought for awhile that it might be caused by water vapor arriving from asteroid impacts of the outer planets. I had to give up on the idea when it was pointed out that it is not something one could expect to occur at anything like regular intervals. I then got the idea that heat from the Earth's core might be stored by the ice sheet until there was sufficient to start melting, something that might happen periodically. Much time was spent calculating the diffusion of heat through ice but it was eventually concluded that there was just not enough heat from below.

Another belief in mainstream climatology is that life processes control the carbon dioxide concentration in the atmosphere. Absorption and decay of vegetation causes annual fluctuation and long-term climate is regulated by storage of $CO_2$ in the form of limestone and coal with eventual release by erosion and volcanic activity. So, with this in mind I came up with the idea that, during a glacial stage, volcanoes and the decay of dying forests would cause a buildup until finally greenhouse warming was sufficient to overcome the feedback cooling of the glaciers and cause them to melt. I mentioned this to someone at Environment Canada and was told "No. There is data that shows $CO_2$ concentration decreases and bottoms out at the end of the glacial stage". "What!!"? I'm afraid I expressed my shock somewhat discourteously because this means that the biomass is not the main source and sink for $CO_2$ and that there is data to prove it. I went looking through the stacks of the Meteorological Service Library, as suggested, and found a wealth if data that is being completely ignored.

Each of the following chapters is centered by a science report, all submitted to journals except #8, some published and some not. It is unusual to resort to a science report format in a textbook, however it is difficult by other means, to demonstrate how ideas and arguments are supported and conclusions arrived at. As any scientist will attest, sticking as close as possible to the prescribed form for science reports, namely: Introduction (what is the problem), Procedure (what you do), Analysis (what does it mean), Discussion (how does it fit with everything else) and Conclusions (usually one sentence for each), will often magically solve the problem. Using this format also helps others to understand and reference your work. Fitting words into a science report is not always easy and does not always work, because of wrong assumptions, missing information, mistakes, difficulty expressing ideas, etc., but it has great advantages over straight prose if problems are complex and if you have to justify your claims. If the text of reports is found to be hard going, perhaps one could concentrate on abstracts, introductions and conclusions. Each report is meant to be self sufficient thus many ideas, figures and arguments are repeated.

# Chapter 1
## Ocean Lag Time

I began my analysis by attempting to fit together the $CO_2$ and atmosphere temperature data obtained from ice core samples from Antarctica. I found that if the temperature scale for the previous interglacial period were set to match the increasing $CO_2$ concentration, at the start of the period, the temperature line would eventually have to fall below the $CO_2$ level. It occurred to me that the ocean temperature might be lagging behind the rest of the environment, thus eventually causing a change in the ability of the atmosphere to retain $CO_2$. It looked like the ocean/atmosphere crossover might be delayed by something like 20,000yr. If the ocean temperature were to be delayed I thought perhaps the atmosphere and ocean might cycle opposite to each other and at 20,000yr there might be an interaction of this cycle with the well-known 22,000yr precession-related climate cycle.

I had found glaciation data in Encyclopaedia Brittanica, for the last 1.8million years. It had been obtained by Quaternary Research using oxygen isotope ratios in ocean sediment cores. So I hastened to look and, sure enough, there seemed to be enhanced glaciation activity at about 350,000yr intervals. As you will see in the upcoming report, 352,000 happens to be exactly divisible by 22,000 and a matching ocean/atmosphere cycle has to have a period of 23,500yr. This gives an ocean lag time of 11,750yr. It was a fortunate mistake I made in thinking that the apparent 20,000yr delay in climate change was the full cycle. If I had realized that a 20,000yr lag would indicate a 40,000yr-ocean cycle I would never have tried to match it with precession.

I present here the unabridged text of the report as published in Theoretical and Applied Climatology.

# 1. World Ocean Temperature Lag Time:
An Analysis Based on Glaciation Data for the Last Two Million Years

## L. G. Bell

*Abstract*

*It is postulated that before the influence of glaciation, it was the amount of cloud cover and the thermal inertia of the ocean that controlled the Earth's temperature. The control system went into oscillation 37myr BP when Antarctica started moving into its present position, the temperature of the ocean and that of the rest of the environment opposing each other in antisymmetric mode. Support for this theory is provided by the observation of periods of enhanced glaciation at regular intervals. The enhancement, being attributed to harmonics with the Earth's 22,000yr-precession and 41,000yr-nutation cycles, allows the calculation of 23,500yr for the period of the ocean/atmosphere-temperature cycle. The corresponding lag time between atmosphere and ocean is 11,750yr.*

## 1 INTRODUCTION

The currently accepted view, e.g. Ericson (1989) is that life and its involvement in the $CO_2$ cycles has controlled the climate over the history of the Earth. It is hereby maintained however, that under normal circumstances, i.e. when not influenced by glaciation, the Earth's climate is controlled by the amount of cloud cover. If there were to be a climate change of short duration, i.e. a few hundred years, as for example a drop in temperature from volcanic activity, clearer skies resulting from an anticyclone over the land would bring the system back to the control point. In addition if there were to be a very gradual change, as effected by an increase in the Sun's heat over millions of years, the amount of cloud cover would adjust itself to allow in only that amount of heat required to keep temperatures more or less the same. Neither short-term nor very long-term climate effects as described would significantly change the ocean temperature. There is a certain type of climate change however that does change the ocean temperature and that sends the self- regulating system out of control.

Climate cycling began perhaps as much as 37 million years before present (myr BP) when Antarctica started moving into its current position. An ice cap formed which made the control system vulnerable and at some point produced just the right amount and duration of climate change for oscillation to begin. It is postulated herein that, because of thermal inertia of the ocean, the temperature of the ocean and that of the rest of the environment began to cycle opposite to each other as depicted in figure 1. The "rest of the environment" includes the land, everything on it, and the atmosphere. For the sake of brevity and because the atmosphere has the main interaction with the ocean the "rest of the environment" is subsequently referred to as the "atmosphere". When the atmosphere is warm the ocean starts heating, but by the time the ocean gets to maximum temperature the atmosphere has cooled to its minimum. To do this, the two environments have to cycle directly opposite, i.e. in antisymmetric mode and be more or less sinusoidal. Since the cycles are sinusoidal the lag time between the two is half the period (Figure 1). Cloud cover now works against control. When the ocean is warm, cloud cover cools the climate (atmosphere), when the ocean is cool, clear skies warm the climate. This tends to maintain the oscillation. Under normal circumstances, in spite of the impetus described, the oscillation would probably die away but with glaciers growing, each cycle is given an extra kick to maintain amplitude and eventually cause an increase.

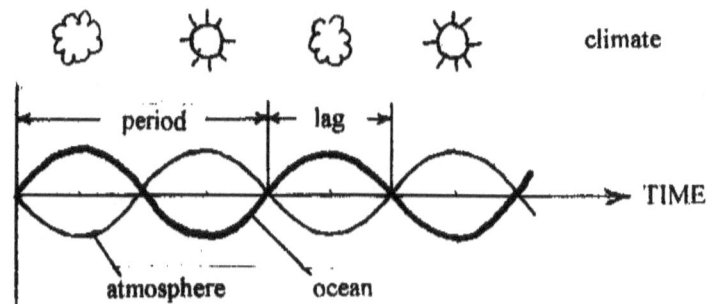

*Figure 1. Stylized depiction of atmosphere and ocean temperature cycles in asymmetric mode.*

The primary objective of the present report is to test this hypothesis against observations. A second objective is to determine the period of the cycles and hence the lag-time between the two environments. (Temperatures of the two environments are effective, relative temperatures rather than actual.)

## 2 PUBLISHED INFORMATION

Following proposed theories that the Earth's climate is affected by its attitude with respect to the Sun, Milutin Milankovic, during the 1920s and 30s, meticulously calculated the Sun's warming ability for various types of cycles, namely: precession (22,000yr), nutation (41,000yr) and orbit (100,000yr). The cycles are described briefly by Ericson (1989) and Schneider (1996). Most experts believe that these cycles, with heat effects as calculated by Milankovic, are what determine the period of glacial cycles (e. g. Schneider, 1996).

Data on the extent of glaciation over the last 2myr were reported by Porter et al (1989) and are shown here, as published in Encyclopaedia Britannica (1997), in figure 2. The extent of glaciation is related to the ratio of oxygen isotopes in ocean sediments. Positive readings produce shaded areas representing glaciation. Along with generally increasing glaciation over the 2myr time period, it will be noted that there are periods of enhanced glaciation at regular intervals averaging about 350,000yr.

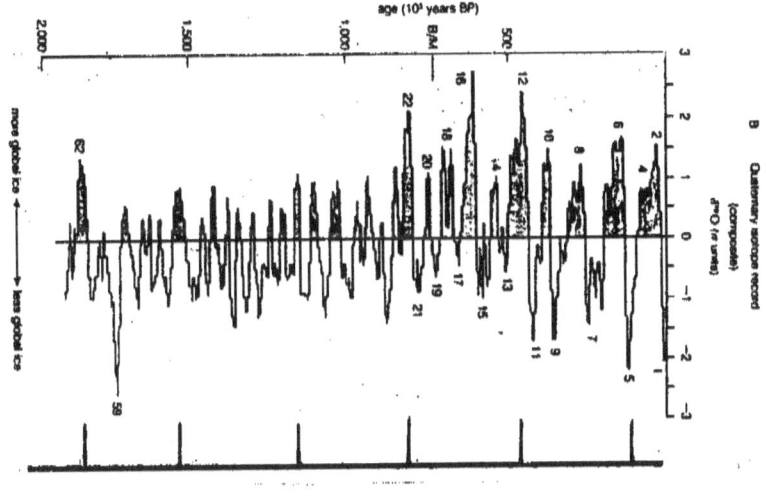

Harmonic glaciation nodes at about 352,000 yr intervals.

*Figure 2. Record of glaciation over previous 2myr as measured by Porter et al, 1989 using the ratio of $^{18}O$ to $^{16}O$ in ocean sediments and showing harmonic nodes at marked intervals.*

## 3 LAG TIME

It is proposed that the periods of enhanced glaciation are nodes of a harmonic between two frequencies and are identified as such in figure

2.  Assuming that this is so, there must be two climate-affecting cycles with periods of similar length that divide evenly into 350,000yr. The 22,000yr-precession cycle, which is known to have an influence on the climate, divides evenly with 16 cycles into 352,000yr. (Note: The data of Porter et al (1989) in figure 2 take the timing of the Brunhes/Matuyama magnetic field reversal to be 720,000yr BP whereas a recent review by Peltier (2001) states that 780,000yr BP is currently accepted as being more accurate. The agreement noted here between the measured value, 350,000yr and calculated, 352,000yr for node period would support an argument that 720,000yr BP is the better value).

The precession cycle is thus assumed to be one of the node-forming cycles and the other to be the ocean/atmosphere temperature cycle. Of the two opposing cycles the atmosphere would be the one most likely to affect glaciation. The atmosphere cycle must have a period of sufficient length to divide evenly, either 17 or 15 times, into the same time period (e.g. Figure 3). Dividing 352,000yr by these numbers gives respective period lengths of 20,700yr and 23,500yr. Half these amounts, namely 10,400yr and 11,750yr, are the would-be respective lag times.

Figure 2 also provides a means of determining which of the two sets of values is correct. It shows what appears to be a larger node at 600,000yr BP, centered between two of the 352,000yr nodes marked 12 and 22 and also centered in a period of about 400,000yr when the oxygen isotope ratio is higher than normal. There is also another, less pronounced but fairly prominent indication at about 1.84myr BP, the two having a separation of 1.24myr. This suggests that there is another even grander harmonic with a period of 2.48myr with mid-node at 1.24myr.

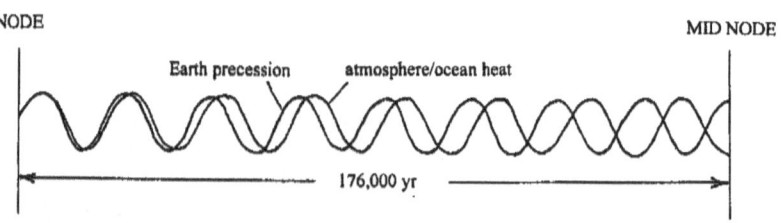

*Figure 3. One half of a 352,000yr-node period for harmonic between the stylized cycles of atmosphere/ocean temperature and 22,000yr precession.*

The formula for a harmonic is $P1(n+1)=P2(n)$, where P1 is the period of the shorter cycle, P2 that of the longer and n is the number of the longer cycles between nodes. Each side of the equation is the node period. For the 352kyr harmonics we have the alternatives:

    A. P2 is 22; n is 16, node length 352 -gives P1 20.7

    B. P1 is 22; n is 15, node length 352 -gives P2 23.5

Only one of these can be valid.

A possible choice for involvement is the nutation cycle (nodding of the Earth's axis towards and away from the Sun at 41,000yr intervals), because half the period at 20.5kyr is close to the period of the other three. Trying each in turn:

    C. P1 is 20.5; P2 is 20.7   - gives n about 100 and node length 2.07myr.

    D. P1 is 20.5; P2 is 23.5  - gives n about 7 and node length 164.5kyr.

    E. P1 is 20.5; P2 is 22      - gives n about 14 and node length 308kyr.

It is not possible for the C harmonic to harmonize with the 352kyr nodes, thus the lower wave length cycle is essentially eliminated from the competition. Now trying the formula with the 352kyr nodes and nodes from D and E.:

    F. P1 is (164.5)2 = 329, P2 is 352 - gives n about 14 and node length 4,928kyr.

    G. P1 is 308; P2 is 352  - gives n exactly 7 and node length 2,464kyr.

The final possible combination is between D and E:

    H. P1 is 308; P2 is 329  - gives n about 15 and node length 4,935kyr.

(A diagram showing the relationship between the various harmonics is shown in figure 4.)

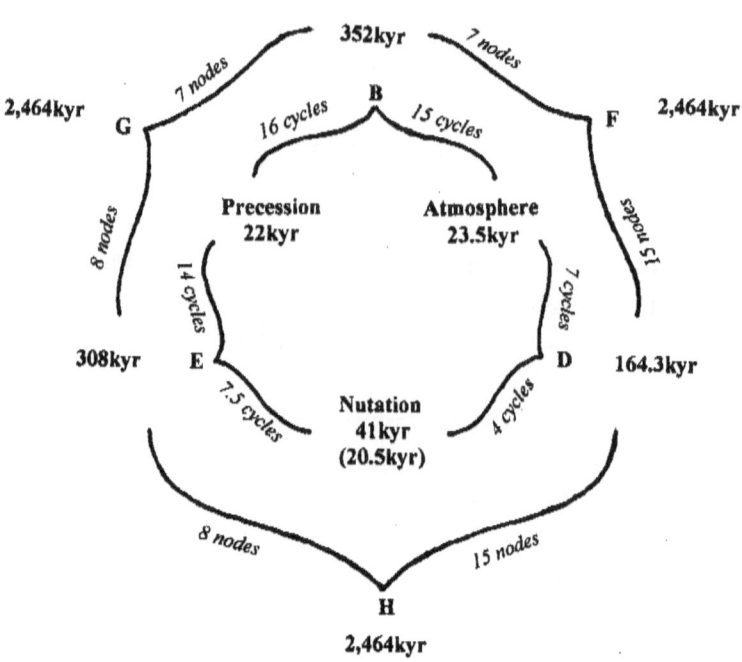

*Figure 4. Diagram showing the various harmonics between precession, nutation and atmosphere cycles. Each bracket represents a harmonic, the inner ring between each pair of cycles and the outer ring between the harmonic nodes of the inner ring. Node lengths sometimes legitimately divided by two.*

Half of 4,928 is 2,464 thus all three harmonics, F, G and H are related. The 23.5kyr-temperature cycle seems to have been created, or at least adjusted to size, by the other two. Note that half of 2.464myr is close to the grand mid-node length of 1.24myr observed in figure 2. Precession and nutation cycles are given in most references to only two significant figures. If it were assumed that 22,000yr is exactly correct for precession, 41,067yr for nutation would make everything work out exactly. The converse of course could also be true i.e. 41,000yr for nutation, or something close to it, being exactly correct and the 22,000yr figure adjusted to something else. The node at 600kyr BP is then a confluence of all three cycles, precession at 22kyr, nutation at 41kyr and ocean/atmosphere at 23.5kyr. The formation of the grand node requires interaction between three distinct cycles, thus the lower-period temperature cycle is definitely eliminated and the larger one, at

23.5kyr, declared the correct choice. (The 100,000yr-orbital cycle does not enter the picture.)

One would expect the grand node to coincide exactly with one of the 352kyr nodes. It probably does not because of a delay in the buildup of glaciation as the node is approached. Because of albedo feedback, glaciation tends to feed on itself. The glacial effect of the grand node is thus pushed forward by 175kyr or more. The reason for glacial enhancement at grand-mid-node is that it coincides with a 352kyr mid-node and a 308kyr node. Data are missing but it seems that the mid-node effect is also delayed by about 175kyr.

## 4 PREFERENTIAL COOLING

Since glacial periods last through many precession or atmosphere-temperature cycles it could be argued that the heating and cooling during the nodal period, though enhanced, once averaged over time, would cancel each other and there would be no noticeable change. However during an enhanced cool period approaching a node when skies are cloudy, the glaciers grow more than normal, then during the ensuing enhanced warm period when skies are clear, the Sun's heat is reflected away by the extra ice. The cool periods are thus more effective in cooling the environment than the warm periods are in heating it. This results in extra glaciation at the nodes.

In the middle of the time between nodes there is a period when cool periods of one cycle are opposite warm periods of the other. One would again expect in this circumstance that the two would cancel each other and not produce any tendency for increased glaciation. However precession, while beginning to cycle in antisymmetric mode with the atmosphere, also begins in symmetric mode with the ocean as the mid-point is approached (Figure 3). During periods when there is a maximum of Sun's heat arriving due to the precession cycle, the ocean is warm, thus there is cloud-cover cooling. On the other hand when the Sun's heat is at its minimum the skies are clear. This reduces the average temperature around mid-nodal points and provides a slightly greater chance of having larger glaciers and extra feedback cooling. Whether or not increased glaciation actually happens at the mid-point, or at the nodes for that matter, depends on the glaciation cycle and also on the proximity of a grand node. The period of glaciation cycles, unlike the enhancement seems, in recent times, to be unaffected by either primary cycles or their harmonics.

## 5 A PERSISTENT VIBRATION

According to the present hypothesis the ocean/atmosphere temperature cycle increased in amplitude over the many millions of years that it took for Antarctica to move into position. Glaciers began to form on Greenland 2.4myr BP (Souches, 1997). Then, as will be described in an upcoming report (Bell) many augmenting effects were introduced. Climate cycling was enhanced by the albedo feedback of increasing ice sheets, by the greater solubility of $CO_2$ in colder seawater and by the effect of ocean temperature on cloud cooling, which in turn determines the $CO_2$ concentration in the atmosphere. All of these factors continued to increase both the period and amplitude of glacial/interglacial cycles right to the present time. In addition the growth and decay of the biomass, though having a moderating effect on the climate, introduces another influence on atmosphere and ocean temperatures.

It is surprising with all this going on that there would still be evidence of the original 23,500yr-ocean/atmosphere cycle. On the other hand a vibrating system will retain a certain note, or notes, in spite of many other tunes that are played (Fowles, 1977). This is apparently what happened and indeed what is still happening. We must think of the Earth's climate as a vibrating system.

## 6. CONCLUSIONS

1. Before cycling began the climate was controlled by the amount of cloud cover.
2. The system went into oscillation with ocean and atmosphere temperatures cycling opposite to each other in antisymmetric mode.
3. A harmonic between the 22,000yr-precession cycle and the atmosphere temperature cycle produces nodes of enhanced glaciation at 352,000yr intervals.
4. A grand node for a harmonic of the two heat cycles, precession and nutation and the atmosphere temperature cycle occurred about 600,000yr BP.
5. The ocean/atmosphere cycle has a period of 23,500yr.

6. The lag time between the ocean temperature and that of the atmosphere is 11,750yr. The atmosphere also lags the ocean by the same amount.

7. Precession and nutation cycles do have an effect on glaciation but they do not determine the period of glacial/interglacial cycles.

Acknowledgments

The author gratefully acknowledges kind permission by *Quaternary Research* to republish data in figure 2. Advice and support from J. C. Anderson and editorial comment by J. A. L. Robertson is also greatly appreciated.

References:

BELL LG, (in press) Ice Age Mystery: A Proposed Theory for the Cause of Climate Change

Encyclopaedia Britannica (1997), 5: 289, and 19: 860-868.

ERICSON J (1989), Living Earth. Blue Ridge Summit PA: Tab Books, 16, 109-111 pp

FOWLES GR (1977), Analytical Mechanics, Third Edition. New York: Holt, Rinehart and Winston, Ch.10: 276p

PELTIER WR (2001), Earth System History, Encyclopedia of Global Environmental Change, Ed MUNN T. Chichester, John Wiley and Sons, 1, 46

PORTER SC (1989), Some Geological Implications of Average Quaternary Glacial Conditions. Quaternary Research. 32:245-261, Figure 1: b and c.

SCHNEIDER SH ed 1996, Encyclopedia of Climate Change and Weather, Oxford University Press, Vol. 2, Milankovitch:507, Vol. 1, Ice Ages:423

SOUCHES R 1997, The buildup of the ice sheet in central Greenland. Journal of Geophysical Research 102(C12):26317-26323

It is reasonable that the ocean temperature should lag changes in temperature of the rest of the environment. Making an analogy with the kitchen range, a large pot of water on a small burner would not be expected to immediately follow the temperature of the heating element. It takes time to respond. For the ocean we have a particularly large pot, small burner. So it takes a long time to heat. It may also be expected that, when the ocean is eventually heated above the rest of the environment, clouds form and this cools the atmosphere. This is a revolutionary idea because it means it is the ocean temperature, rather than the generally-accepted variation in Sun's heat as calculated by Milanchovic, that determines the periods of glaciation cycling.

# Chapter 2
## Ice Age Mystery

As mentioned, the original objective of the program was to determine the mechanism by which the Ice Age came to an end, when feedback cooling from the ice should have made it next to impossible. The second report in the series deals with this issue. It is pasted here as published in Theoretical and Applied Climatology and will be referred to as the Ice Age report.

It is maintained in climatology theory that the biomass is the main source and sink for $CO_2$. It is easy to refute this because it is also claimed that the buildup of $CO_2$ in the atmosphere at the start of interglacial periods is the result of "radiative forcing". An increase in the Sun's energy supposedly causes $CO_2$ to come out of the ocean, which means indeed that the ocean would at least have to be the main source. If the ocean is the main source it also has to be the main sink. The source and sink is therefore the ocean, not the biomass. Jacob tries defining radiative forcing on page 133 of "Introduction to Atmospheric

Chemistry". "*The radiative forcing caused by a change delta m in the atmospheric mass of a greenhouse gas X is defined as the resulting flux imbalance in the radiative budget for the Earth system*". It is difficult to imagine exactly what this means (he seems to say that sunlight releases $CO_2$). It was accepted as truth at the time of writing the report and the temperature and $CO_2$ curves from Antarctica were adjusted so that the two would increase in concert. This caused some problems that had to be straightened out in later reports.

# 2. Ice Age Mystery:
# A Proposed Theory for the Cause of Long Term Climate Change

## L. G. Bell

*Abstract*

*Analysis of available data shows that the duration of the glacial/ interglacial cycle is determined by the time for the ocean to go through one major temperature cycle. At the start of an interglacial period, clear skies with consequent release of $CO_2$ from the ocean, warm the atmosphere, which in turn eventually warms the ocean to its maximum. Cloudy skies then cause the climate (land and air temperature) to cool and the $CO_2$ to be reabsorbed to start glaciation preliminaries. The albedo feedback effect of the glacial ice, a relatively warm ocean, which produces enhanced cloud cover, and the increased solubility of $CO_2$ in cold seawater ensure a long period of glaciation. Glacial periods end when pack ice spreads out on the ocean cooling it until reduced cloud cover once again allows the Sun's heat, un-reflected by cloud cover, to melt the ice and release $CO_2$ back into the atmosphere.*

## 1 INTRODUCTION

The ending of the Ice Age has long been an unsolved mystery. Starting about 70,000yr BP, ice sheets had spread to the point that they covered most of the northern half of North America and much of northern Eurasia. Ice reflects the Sun's heat back into space. Continued growth of the glaciers is therefore expected to have been ensured by what is called albedo feedback cooling; the colder it gets, the more snow, the more heat lost, the colder it gets. Common logic dictates that it should have survived any conceivable change; and yet, rather suddenly about 15,000yr ago, the Ice Age came to an end.

The Ice Age is the last glacial stage of the Pleistocene glacial epoch, sometimes referred to as the Great Ice Age, a long history of intermittent glaciation with the duration of glacial and interglacial stages generally increasing over time. Glacial stages thus ended many times, and presumably for the same reasons. The investigation of the final glacial retreat, therefore, has had to take this longer history into account. As a result, as will become apparent, there is much more to be learned than the solution to the mystery. As stated by Peltier (2001) in his recent review, "it is important that we begin to understand the changes to the environment that humankind has begun to induce. This requires an understanding of the obvious drastic changes that have occurred naturally throughout history". The Great Ice Age is of particular interest in this regard.

The theory of intermittent glaciation most commonly accepted is based on the Earth's attitude with respect to the Sun. The Sun's warming ability was calculated in the 1920s and 30s by Milankovic for various types of cycles, namely: precession (wobbling of the Earth's axis, 22,000yr), nutation (nodding of the axis towards and away from the Sun, 41,000yr) and orbit (changing from a circle to an ellipse, 100,000yr). Ericson (1989) and Schneider (1996) describe the cycles briefly. Because of the timing of glacial stages in the recent past, most climatologists believe that a combination of the attitude cycles is what determines the period of glacial/interglacial cycles (Schneider, 1996). A weakness of this theory is that no one has yet explained how these cycles of constant period can produce glacial cycle periods that are known to change with time. It was concluded in a previous report (Bell, 2002a) that precession and nutation cycles do have an influence on glaciation but they do not control the cycle period. Arguments in (Bell, 2002b) will rule out the involvement of any combination of the Earth attitude cycles.

A new theory for glacial cycles is hereby proposed based on information recently available. The premise is that the glacial cycle period is determined by the time required for the ocean to go through one major heating/cooling cycle. Note that temperatures referred to are the effective, rather than actual, relative temperatures of the ocean and the rest of the environment. The "rest of the environment" includes the land and atmosphere and is referred to herein as the atmosphere. The temperature of the "atmosphere" or the amount of cloud cover is sometimes referred to as the "climate".

## 2 PUBLISHED INFORMATION

### 2.1 Glaciation Data

Data on the extent of glaciation over the last 2myr were reported by Porter (1989) and are shown here, as published in Encyclopaedia Britannica (1997), in figure 1. Profiles were deduced from the measurements of the ratio of $^{18}$O to $^{16}$O in ocean sediments. There are two sorting mechanisms for the isotopes, one determined by the freezing of water on the glacier and the other by the temperature of the ocean surface layers at the time of deposition. In general, increasing positive readings to the left indicate more glaciation to define glacial stages, but higher positive readings can at times, also indicate a lower ocean temperature. Negative readings to the right indicate less glaciation to define interglacial stages, but can also, supposedly, indicate a higher ocean temperature.

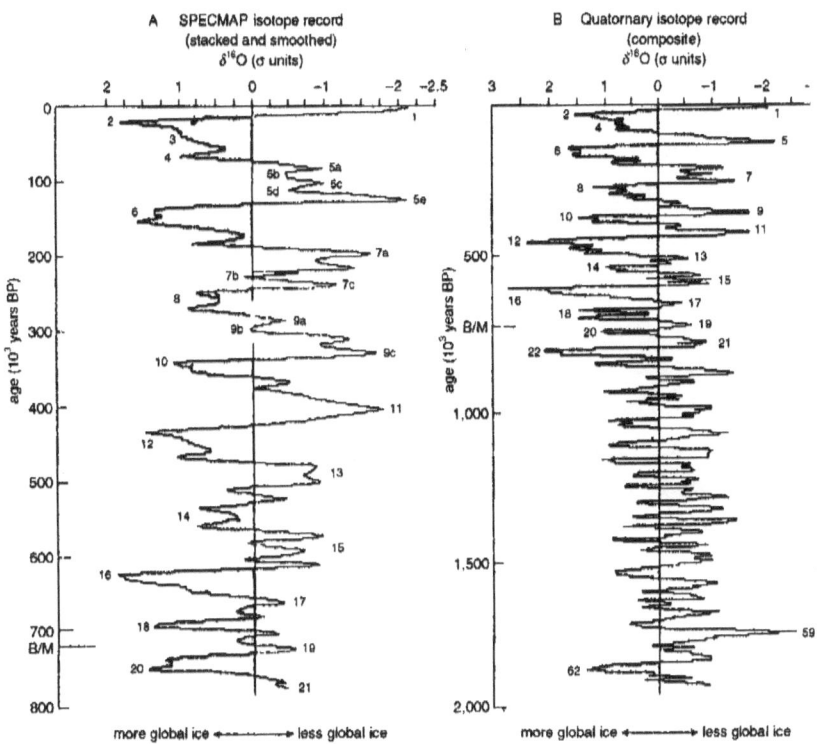

*Figure 1. Data presented by Porter (1989) of the ratio of $^{18}$O to $^{16}$O in ocean sediments and limestone deposits, each plot averaged for several locations. The data mainly indicates the degree of glaciation over time thus giving an idea of the relative size and timing of glacial and interglacial periods. Occasionally changes may be attributed to changes in ocean temperature. The Brunhes/Matuyama reversal, used to set the time scale, is a reversal of the Earth's magnetic field at 720,000yr BP.*

The B-side on the right of figure 1 shows the longer history (2myr) of cycles with generally increasing period and amplitude before as well as after the onset of major glaciation about 750,000yr BP. The glaciation data show unexpected erratic behavior but it is noticeable that the duration of the warm part of the glacial/interglacial cycle stays relatively constant over time, whereas that of the cold part increases by a factor of perhaps six.

The shorter history on the A-side of figure 1 shows averaged data taken from several sites, different from those used for the B-side. The shorter-term data provide a magnified view of the last 800,000yr. After about 500,000yr BP, glacial stages, as determined by a positive oxygen isotope ratio, occur at intervals of approximately 100,000yr. During the glacial stages there are what appear to be periodic deglaciations. However if this feature is ignored, glaciation generally increases rapidly in the early stages of glacial periods, then the rate gradually slows down. Toward the end of the glacial period the rate usually increases again, up to a maximum, then drops.

## 2.2 $CO_2$ Data

Barnola et al (1994) measured atmospheric $CO_2$ concentrations for the last 160,000yr from ice cores recovered at Vostok, East Antarctica. Their data, with measurements actually made and published in 1987, are shown here in figure 2. Glacial periods, as determined by the positive oxygen ratios in figure 1A, are shown in figure 2 as discontinuous sections of a horizontal line.

The previous interglacial period lends itself better to analysis than does the present one. To start, the $CO_2$ concentration increased for about 10,000yr to peak at 290 ppmv. For the first 20,000yr, after the initial peak, the concentration remained constant at 270 ppmv then there was a large step-down followed by a slow decrease to about 220 ppmv. After some erratic cycling with the average concentration gradually decreasing, both before and during the glacial stage, there was a sudden rapid increase after the concentration had been reduced to 190 ppmv.

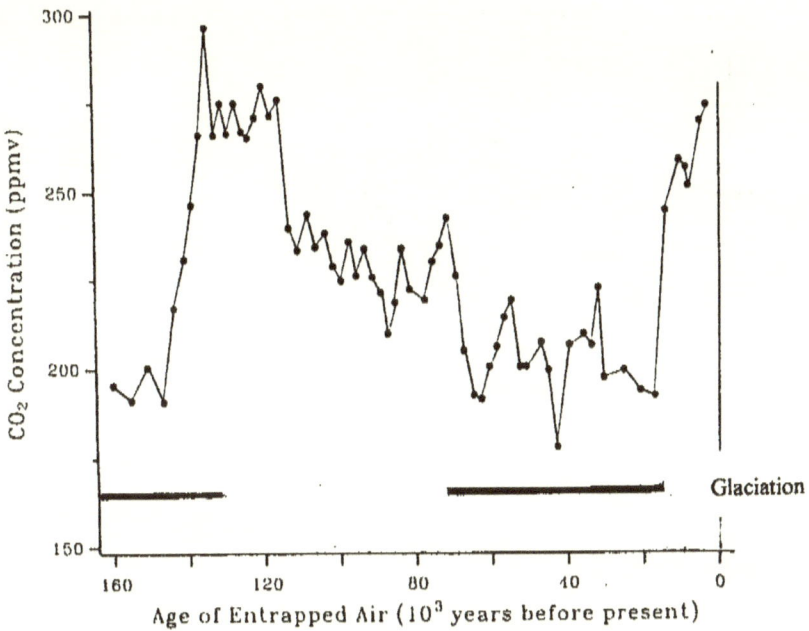

*Figure 2. Atmospheric $CO_2$ concentrations derived from the Vostok ice core (Barnola et al, 1994) for the last major transition. Glacial stages as derived from oxygen isotope ratios in ocean sediments (Figure 1) are shown as a line.*

The large apparent discontinuity at the step-down can be explained by means of subsequent measurements made by Barnola et al (1991). They measured the $CO_2$ concentration profile more accurately and found the discontinuity to be somewhat less severe. These more recent data covering the early part of the previous interglacial period are shown in figure 3. The time scale is reversed from that in figure 2. Starting from the initial peak, there is now a second modest peak of 270 ppmv at 10,000yr, then a gradual decrease in $CO_2$ to 260 ppmv at 20,000yr. This is followed by a more rapid but continuous decrease of 30 ppmv over the next 10,000yr.

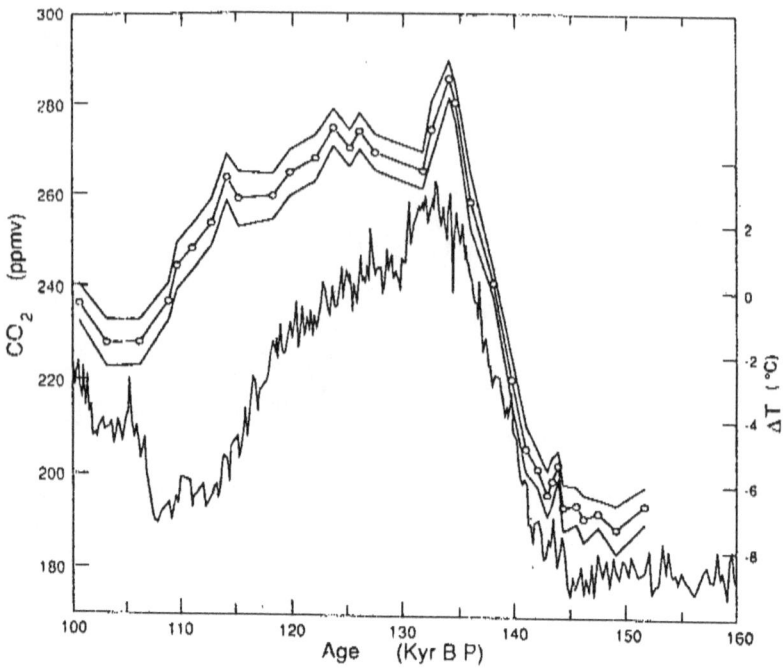

*Figure 3. Temperature and CO₂ concentration profiles for the start of the last interglacial period. Antarctic CO₂ variations from Barnola et al (1991) and surface temperature variations adapted from Jouzel et al (1987) recorded on the Vostok ice core. The CO₂ data shows the envelope of measurement uncertainty.*

### 2.3 Temperature Data

Figure 3 gives ice surface temperature as well as $CO_2$ measurements for the short period of glaciation history at the start of the previous interglacial period. Jouzel et al (1987) obtained these comparative temperature data for the ice surface on Antarctica using hydrogen isotope measurements. Data were actually obtained for the whole 160,000yr of the last major transition and published in Encyclopedia Britannica (1997). The data are modified and used in figures 4, 5 and 6 of this report.

### 3 THE START OF CYCLING

After the end of the Cretaceous period 65million years (myr) before present (BP), the climate was fairly warm and the temperature constant for at least 22myr (Hecht, 1985). It was inferred in Bell (2002a) that during this 22myr period the temperature was controlled by the amount of cloud cover. Instability could have occurred as early as 37myr BP when Antarctica started to move into its position at the South Pole. The cloud-control system went into oscillation with the temperature of the ocean and that of the rest of the environment cycling opposite to each other. The period of ocean temperature cycling in the early stages was determined in the report to be 23,500yr, which means a minimum lag time between the ocean and the rest of the environment, one to the other, of about 12,000yr.

In the early stages, climate cycles were enhanced by a feedback mechanism involving the cloud control system. When land areas were relatively warm, for example, and skies therefore relatively clear, more heat was provided to make the skies even clearer. Once glaciation started, atmosphere circulation patterns would have been affected, especially over the land and this would have affected the amount of cloud cover regionally, but perhaps not globally. The increasing instability of the cloud control system is therefore assumed to have carried on unabated.

## 4 MAJOR GLACIATION

The amplitude, and eventually the period, of temperature cycles began to increase as Antarctica moved into position and as Greenland also moved north to occupy its current location about 2.4myr BP (Souches, 1997). Data in figure 1B only go back as far as 2myr and there was by that time already a significant amount of glaciation occurring. The extent of glaciation subsequently generally increased with time as would be expected with an oscillating system. Serious glaciation started about 750,000yr BP.

The glaciation data particularly figure 1B, show unexpected scatter. The glacial stages do not increase gradually in amplitude, or in spacing, as one would expect. For one thing periods of enhanced glaciation occur at constant intervals. In Bell (2002a) it was explained that this is the result of harmonics between the Earth attitude cycles and the ocean/atmosphere temperature cycle. For another thing glacial periods appear to show considerable variation in size and spacing with some even appearing to have deglaciation occurring halfway through the expected glacial stage. A reason for the scatter and occasional apparent early retreat is believed to be the result of a previously unknown attitude cycle, which will be investigated in Bell (2002b).

## 5 INTERGLACIAL PERIOD

As part of our hypothesis, cloud cover is taken to be the means by which the Earth's temperature is normally controlled. Once oscillation starts, cloud-cover, rather than controlling, now adds to the instability. One of the ways it does this is to effect changes in $CO_2$ level. The present report is based on the premise that a relatively warm ocean is associated with more cloud cover, and a relatively cool ocean with less. Another factor is that as seawater is warmed, the solubility for $CO_2$ decreases (Masterton et al, 1977). With this information it is reasonable to assume that, even with a cold ocean, sunlight heating of a thin-surface layer would cause $CO_2$ to be released to the atmosphere. If the skies are clear (as a result of the ocean being relatively cold) there is more surface heating and $CO_2$ comes out of the ocean. If the skies are cloudy (the ocean relatively warm) there is less surface heating and $CO_2$ is reabsorbed.

The following scenario seems to best fit the $CO_2$ data: At the start of the interglacial period, $CO_2$ continues to come from the ocean because of the clear-climate conditions. It eventually reaches a point at about 10,000yr into the interglacial period where it is being consumed by increasing biomass at such a rate that it produces the first peak, then allows the level to rise to 270 ppmv at 20,000yr. At this point the ocean has warmed sufficiently that cloudy conditions prevail and when averaged globally, $CO_2$ is now being absorbed. This causes the $CO_2$ to gradually decrease to 260 ppmv at 30,000yr into the interglacial period. By then the ocean has already started to cool, and with cloudy conditions continuing, it rapidly absorbs $CO_2$ over the ensuing 10,000yr period reducing the concentration by another 30 ppmv to start glaciation preliminaries. From the start of the temperature increase the whole process has taken about 40,000yr.

The data comparing surface temperature and $CO_2$ profiles in Barnola's second report (Barnola et al, 1991), shown here in figure 3 were re-plotted as shown in figure 4. The time scale was reversed and the temperature scale increased by 22% and raised to maintain a constant cloud-factor contribution for the increasing-$CO_2$ part of the curve. (22% was chosen because it gives the best fit.) Since we are usually speaking of relative temperatures this seems an acceptable thing to do. It appears from this that global temperatures do increase in step with $CO_2$ concentration. This idea cannot as yet be fully accepted because, as reported by Barnola et al (1991), the temperature increase precedes the $CO_2$ increase.

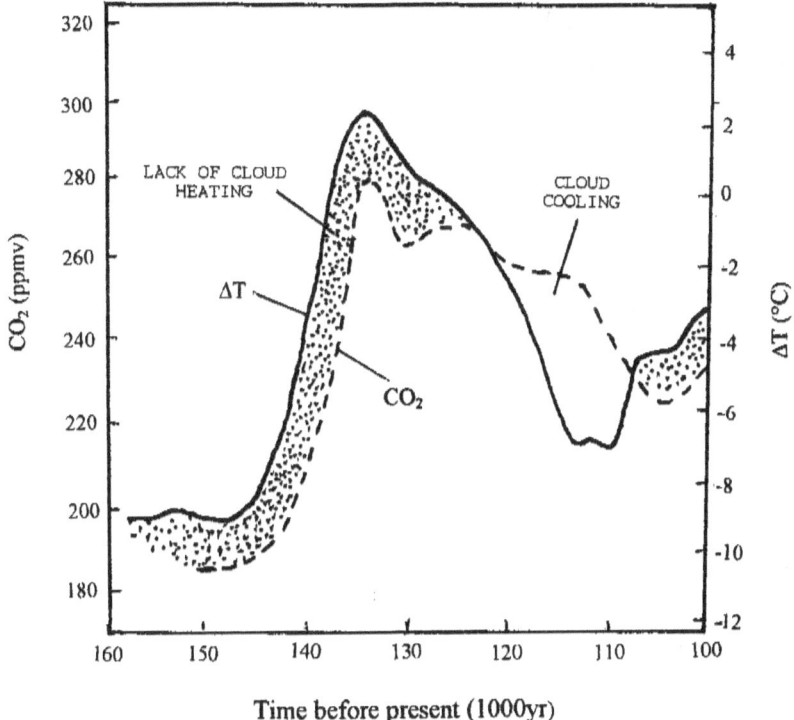

*Figure 4. Variations of Antarctic surface temperature and atmospheric $CO_2$ concentration adapted from figure 2-3. The time scale is reversed and the temperature scale raised and increased by 22%.*

On figure 4 the shaded areas between curves represent, for the most part, extra warming caused by the clear skies that are the result of the ocean being relatively cold. This remains a factor in providing heat for the first 20,000yr into the interglacial period. (Part of the contribution at the top of the curve might be greenhouse warming from transpiration $H_2O$ as $CO_2$ is consumed by the biomass (Schneider, 1996 and Veizer, 2001)). The unshaded area between the curves represents cloud-cover cooling that is the result of the ocean being relatively warm. Greenhouse warming is apparently ineffective if the skies are cloudy since the temperature line falls well below the $CO_2$ line during this time. The cloud cover of this period between about 122,000 and 108,000yr BP, along with consequent absorption of $CO_2$, cools the environment and sets the stage for the next glacial period. Cloud cover thus plays a major role to both start and end each glacial stage. Our experience is that cloud cover changes minute by minute. However, as claimed by Warren et al (1986), global cloud cover is something that can be measured accurately, and that changes only gradually over long periods of time.

It will be noted that it is not possible to adjust the temperature curve in figure 4 so as to have a cloud-cover factor uniform for the first part of the curve. The contribution is twice as much on the sloped portion as it is at either end. The surface temperature measurements on Antarctica have been assumed to represent climate conditions of the whole planet but perhaps skies clear at the South Pole sooner than elsewhere. This would be the result of the polar anticyclone and would explain why measurements show a temperature increase before an increase in $CO_2$. To have the curves in figure 4 maintain a constant cloud-cover factor requires moving the temperature line forward by 1500yr. This would be the lead-time of Antarctic surface temperature over that of the tropics, where most of the $CO_2$ is released. On the other hand during periods of rapid cloud-cover cooling Antarctica would lag behind the atmosphere because clearer skies in the polar region would keep it relatively warm. With this in mind the curves of figure 4 were re-plotted, as shown in figure 5, with temperature curves representing the atmosphere adjusted by up to a maximum of 1500yr depending on the rate of change, forward on heating and backward on cooling.

Modification of the temperature profile, as in figure 5, solves many problems with the data. Temperature and $CO_2$ now increase at the same time. There is a much better fit of atmosphere temperature with the $CO_2$ data, that is to say a uniform cloud-cover factor is now maintained for the heat-up curve. The curve through the maximum at 130,000yr BP and the minimum at 110,000yr BP also looks much more reasonable and an ocean temperature curve with lag time of 12,000yr now fits perfectly for absorption and release of $CO_2$. This adaptation adds strength to the finding, from glaciation data in Bell (2002a), that the ocean temperature lags the rest of the environment by 12,000yr. It supports the ideas that a relatively warm ocean produces more cloud and that the amount of cloud cover is what controls $CO_2$ absorption and release. It also supports the idea that when skies are clear atmosphere temperature increases in step with $CO_2$ concentration.

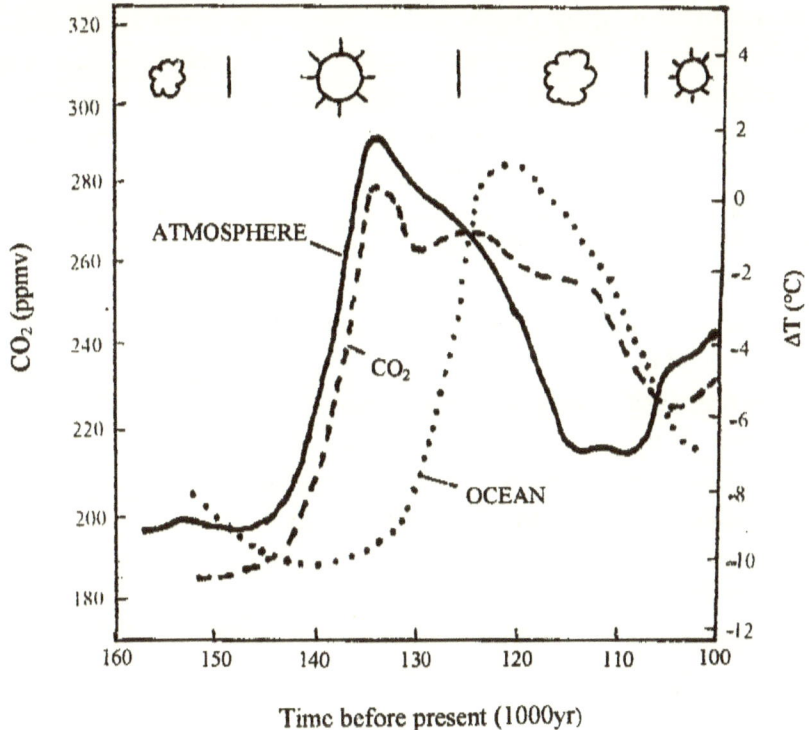

*Figure 5. $CO_2$ concentration and temperature profiles. Surface temperature from figure 4 has been modified so as to represent the global atmosphere temperature. The ocean temperature is projected from atmosphere and resulting expected climate conditions are indicated.*

During the latter part of an interglacial period, before major glaciation begins, the $CO_2$ concentration begins to cycle erratically. There are alternative explanations for this. Most likely the $CO_2$ concentration increases when the climate is clear and decreases when it is cloudy. This in turn drives the atmosphere temperature up and down and the ocean lags behind causing erratically alternating clear and cloudy conditions, which causes the $CO_2$ concentration to cycle erratically. The average $CO_2$ concentration continues to decline as the climate becomes cooler, the $CO_2$ being absorbed by the ocean because of the predominance of cloudy conditions.

## 6 GLACIAL PERIOD

Once glaciation starts, the atmosphere temperature drops dramatically because of cooling by the ice fields. It then recovers somewhat as shown on figure 6. Atmosphere temperature is usually determined by something else, such as $CO_2$ concentration or cloud cover, but perhaps this is one occasion when it follows the ocean temperature that peaked about 15,000yr earlier. There is a short period of clear skies but, with the ocean marginally warmer for most of the glacial period, cloudy skies cause the $CO_2$ concentration to continually decline. Albedo feedback cooling by the ice fields, cloudy conditions brought on by a relatively warm ocean and the increased solubility of $CO_2$ in cold seawater ensure a long period of glaciation.

During a glacial stage the oxygen ratio should be expected to increase continuously up to a maximum as the ice fields grow but this is usually not the case. During the last two glacial stages for example, the glaciation data of figure 1A show an apparent early retreat signified by a drop in oxygen isotope ratio. The start of the effect is numbered 4 on the Ice Age profile, whereas on the previous glacial stage it is not identified. This unusual phenomenon is to be discussed in Bell (2002b).

## 7 END OF GLACIATION

Several glacial stages show a rapid increase in oxygen isotope ratio toward the end of the glacial stage. For the Ice Age the increase starts about 27,000yr BP (point 3 in figure 1A). As explained in Encyclopoedia Brittanica (1997) it is believed that, as the ocean cools, microorganisms preferentially select $^{18}O$ for their shells, which causes the ratio in sediments to increase. It is proposed therefore that the increase at point 3 was not caused by increasing glaciation, as was the previous gradual change, but by actual cooling of the ocean itself. It is presumed that the climate was cold enough to allow the ocean to start freezing. Pack ice spread out from the land and, with the Sun's heat being reflected away by this ice, the ocean cooled at an increasing rate.

A peak, and sudden drop, in oxygen-ratio data in figure 1A occurs at about 22,000yr BP (point 2). This would suggest that the ice sheets were starting to melt. Arguments will be made however, in Bell (2002b) that melting did not start until the oxygen ratio had dropped to zero at about 15,000yr BP. Activity of the microorganisms preferentially absorbing $^{18}O$ may be reduced when the ocean becomes too cold or freezes and this is what caused the change at 22,000yr BP. The glaciers did not start to melt, nor did the ocean stop cooling and warm up quickly, as indicated by the data. Rather, it is suggested that the ocean was continuously cooling right through this time and rather suddenly the climate changed from cloudy conditions to clear skies as the ocean temperature crossed the atmosphere-temperature line.

The Earth is now about 60% covered by clouds, (52% on land and 64% on ocean, (Warren et al, 1986&8 resp.)), which have about the same ability to change the Earth's albedo, as do glaciers (Encyclopaedia Britannica, 1997). One can imagine that a significant change in this current amount of cloud cover would easily overcome the albedo effect of the glaciers. Less evaporation and the relative ocean/atmosphere-temperature effect, greatly reduce cloud cover as the ocean cools. It is hereby proposed that the resulting change in cloud cover is what starts the glaciers melting and eventually releases $CO_2$ from the ocean.

The finding that $CO_2$ was released at an increasing rate during a time that melt-water was still cooling the ocean and the $CO_2$ buildup was warming the atmosphere seems to go against conventional wisdom. It is normally assumed that the $CO_2$ would go the other way. The suggestion is that it is the increased sunlight that releases the $CO_2$. At all times $CO_2$ is released to the atmosphere at low latitudes and reabsorbed at high latitudes. Unhindered sunlight would tend to heat up the surface of the ocean especially at the equator and thus upset the balance in favor of a buildup in the atmosphere.

From arguments above it is apparent that the ocean goes through one major heating/cooling cycle for each glacial cycle. The first observation leading to this conclusion was that warm periods vary less in duration than do cool periods (figure 1B). This might be explained as follows: The duration of warm, interglacial periods is mainly determined by the time required to heat and then cool the atmosphere. As the temperature swings get larger, more, and then less heat, is provided by the disparity between ocean and atmosphere temperatures. The atmosphere heats up faster and cools down faster. This tends to keep the duration of interglacial periods more or less constant. Glacial stages on the other hand get longer because, as cycles increase, the ocean has to cool more to overcome the albedo feedback cooling of the extra ice. Glacial stages only end when the ocean temperature gets below that of the atmosphere. Since that is what is required, it is the ocean temperature cycle that determines the period of the glacial/interglacial cycle.

As the glacial stage is ending, the glaciers are changed by the sudden warm conditions from "polar" or cold, to "temperate" or warm, as described by Sharp (1988). This makes the ice much more fluid with the result that the glaciers advance and retreat as they melt. The rate of melting is increased by this mechanism and there is also an increase in the number of icebergs being calved into the ocean. The extra floating ice cools the ocean even more as the rest of the environment is warmed, which increases temperature disparity and helps to keep the system oscillating.

## 8 TEMPERATURE PROFILE

An attempt is made to estimate the ocean temperature profile for the last major transition by projecting from the profile for the atmosphere. The atmosphere temperature was taken to be that of the ice surface on Antarctica as determined by Jouzel et al (1987) and published in Encyclopedia Britannica (1997), modified by the lead and lag between Antarctic surface and the planet as a whole during heating and cooling as discussed above. The two profiles are shown in figure 6. The ocean is assumed to lag the atmosphere by about 12,000yr. Besides lagging behind the atmosphere, the ocean temperature is assumed to be slightly less variable but otherwise only affected by freezing and glacial runoff. Transition events from figures 1 and 2, as well as from Bell (2002b), are also marked on figure 6. Considering the problems with measurement there is a reasonably good match of time scales for ocean-sediment data (Porter, 1989) and ice-core data (Barnola et al, 1994 and Jouzel et al, 1987), however by 150,000yr BP the discrepancy is about 15,000yr.

The only thing that definitely does not fit the theories described so far is the apparent deglaciation in the Ice Age glacial stage starting at 68,000yr BP. This "deglaciation" was mentioned above in section 6 and is found here to occur, not at an atmosphere temperature maximum, as one would expect, but at a minimum, and also when the climate is thought to have been cloudy.

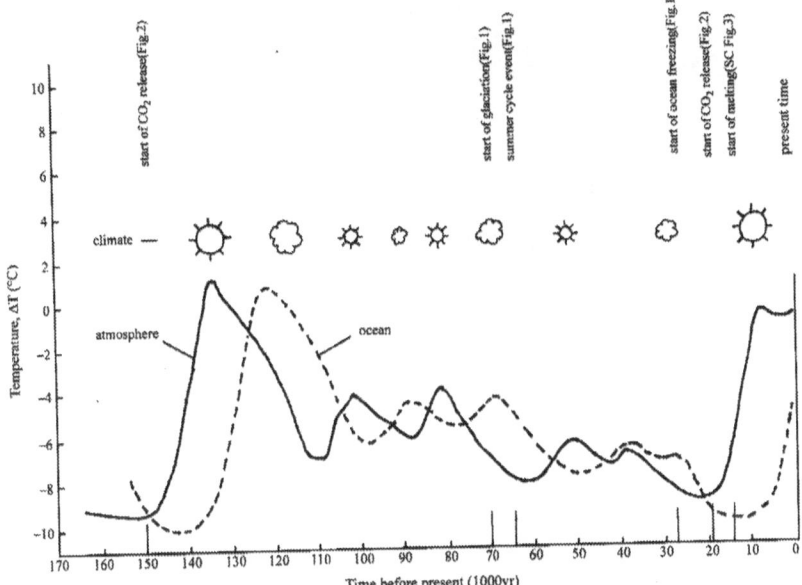

*Figure 6. Temperature profiles for the ice surface of Antarctica from Jouzel et al, 1987 (modified to represent atmosphere) and that projected for the ocean for the last 160,000yr. Expected climate conditions are shown as well as certain transitional features from figures 1 and 2.*

## 9 PROPOSED SEQUENCE OF EVENTS AND CONCLUSIONS

1. Cloud cover, rather than greenhouse gas concentration, is what regulated the Earth's climate over its longer history.

2. When land areas moved near the poles, cloud-cover control started oscillating causing the ocean temperature to cycle opposite to that of the rest of the environment. Increasing glaciation then caused continued growth of the cycles.

3. During these early and later major climatic changes the amount of cloud cover is determined by the ocean temperature relative to that of the rest of the environment, a relatively high ocean temperature producing more clouds and cooler climate, a lower ocean temperature less clouds and warmer climate.

4. For recent major cycles, Antarctic surface temperatures lead those of the rest of the environment during periods of heating and lag on cooling, in each case by up to 1500yr.

5. A $CO_2$-concentration cycle adds to the thermal cycles. When the tropics warm, extra sunlight releases $CO_2$ from the ocean. Increasing $CO_2$ along with unhindered sunlight warms the environment to start the interglacial period. Continued $CO_2$ buildup is halted and the concentration subsequently kept more or less constant by the intervention of increasing biomass.

6. Once the ocean warms sufficiently, the climate becomes cold and cloudy and the $CO_2$ buildup is largely reabsorbed.

7. When skies are clear, land (atmosphere) temperatures increase in tandem with increasing atmospheric $CO_2$ concentration. On the other hand when skies are cloudy, global temperatures fall well below that expected from the $CO_2$ levels that are extant.

8. The ocean temperature lags that of the rest of the environment by about 12,000yr even for the large cycles at the start of the interglacial period.

9. During the latter part of the interglacial period, and the glacial period as well, temperatures and $CO_2$ concentration exhibit relatively minor cycling, and because of the predominance of marginally cloudy conditions the average $CO_2$ concentration gradually declines.

10. The combination of cloud-cover cooling and decreasing $CO_2$ concentration eventually triggers the next glacial stage.

11. During glacial periods a cool climate is maintained by the albedo feedback effect of the ice sheets, by the predominance of marginally cloudy conditions and by continuing reduction in atmospheric carbon dioxide.

12. Glacial-stage ending is initiated by the ocean starting to freeze. Pack ice spreads out from shore and heat

reflected away by this ice causes the ocean to cool. Eventually this results in a change from cloudy to clear conditions as the ocean temperature drops below that of the atmosphere. The increased sunlight, decreasing albedo as the ice sheets melt, a change to plastic ice flow, and release of $CO_2$, all contribute to ending the glacial regime.

13. The duration of a glacial/interglacial cycle is not related to Earth-attitude cycles as previously thought, but rather to the duration of the major heating/cooling cycle of the ocean, which generally gets longer with time.

14. Extenuation of the data for $CO_2$ concentration and atmosphere temperature suggests that these factors should have remained more or less constant for the last 10,000yr, right to the present time. Ocean temperature, on the other hand, is theorized to have been increasing at a constant rate for the last 5,000yr.

Acknowledgments

Acknowledgment is hereby given for kind permission by *QUATERNARY RESEARCH* to republish figure 1, by *Nature* for figure 2 and by *Tellus* for figure 3. Thanks are expressed for advice and encouragement offered by John C. Anderson. Editorial comments by J. A. L. Robertson and A. Sawatzky are also greatly appreciated.

References:

Barnola JM, Pimienta P, Raynaud D, Korotkevich YS (1991) $CO_2$–climate relationship as deduced from the Vostok ice core: a re-examination based on new measurements and on a re-evaluation of the air dating. Tellus 43B: 83-90.

Barnola JM, Raynaud D, Lorius C, Korotkevitch YS (1994) Historical Record from the Vostok Ice Core. Trends '93A: 7-10, see Boden TA .: 8

Bell LG (2002a) World Ocean Temperature Lag Time: An Analysis Based on Glaciation Data for the Last Two Million Years

Bell LG (2002b) Summer Cycle: Another Earth Attitude Cycle Affecting Climate

Boden TA, Kaiser DP, Sepanski RJ, Voss FW (eds.) (1994) Trends '93A, Compendium of Data on Global Change. ORNL/CDAIC-65, Carbon Dioxide Information Analysis Center, Oak Ridge National Laboratory, Oak Ridge, Tennessee, USA.

Encyclopaedia Britannica (1997) 5: 289, and 19: 860-868.

Ericson J (1989) Living Earth. Blue Ridge Summit PA: Tab Books, 109-111.

Hecht AD (1985) Paleoclimate Analysis and Modeling, Toronto: John Wiley and Sons. 402-403.

Jouzel J, Lorius C, Petit JR, Grenthon C, Barkov NI, Katliokov VM, Petrov VM (1987) Vostok ice core: a continuous isotopic temperature record over the last climatic cycle (160,000 years). Nature 329: 403-408.

Masterton WL, Slowinski EJ (1997) Chemical Principles. Philadelphia: W. B. Saunders Co. 292.

Peltier WR (2001) Earth System History, Encyclopedia of Global Environmental Change, Munn RE ed. Chichester: John Wiley and Sons, 1: 31.

Porter SC (1989) Some Geological Implications of Average Quaternary Glacial Conditions. Quaternary Research. 32: 245-261, Figure 1: b and c.

Schneider SH ed (1996) Encyclopedia of Climate Change and Weather, Oxford University Press: 2 Milankovitch: 507, 2 Water vapor: 828-829.

Sharp RP (1988) Living Ice. Press Syndicate of the University of Cambridge: 26-29.

Souches R (1997) The buildup of the ice sheet in central Greenland. Journal of Geophysical Research, DC. 102(C12): 26317-26323.

Veizer J (2001) University of Ottawa, Private communication.

Warren SG, Hahn CJ, London J, Michervin R, Jenn RL (1986) land and (1988) water. Global distribution of total cloud cover and cloud type amounts over land, DOE/ER/60085 -H1: 23 and over water, DOE/ER-0406: 37.

If you didn't manage to wade through that, the solution to the puzzle is that it was so cold for so long during the Ice Age that the ocean began to freeze. This changed the skies from cloudy to clear, the atmosphere started to warm and the ice to melt. Significant findings of the Ice Age report, among many, are that it is clearing skies that caused release of $CO_2$ to the atmosphere and it was necessary to have the increasing biomass intervene to stop the $CO_2$ increase. With intervention the ocean temperature has time to catch up, which changes skies from clear back to cloudy and $CO_2$ control from release to absorption. A detailed analysis of what is apparently happening is undertaken in Chapter 4.

# Chapter 3
# Summer Cycle

As mentioned in the Ice Age report there was a feature that could not be explained. There appeared to be an adverse effect on glaciation at intervals of about 30,000yr and particular instances of this occurring during the last two glaciation stages, 120,000yr apart. It seemed unreasonable that deglaciation could suddenly start to occur in the middle of a cold period. Consideration of this led to the realization that the oxygen isotope ratio data is a measure of the *rate* of glaciation. Researchers doing the measurements must have realized this and set the zero so that glaciation would usually approximately equal deglaciation. The realization here, that it is a rate measurement, led to an explanation for the negative effect at 30,000yr intervals. The measurement does not record deglaciation during the glacial stages but merely a reduction in the rate of ice accumulation. It was decided that this must be due to a periodic change in heat input.

The discovery of a previously unknown climate cycle is, as one can imagine, quite exciting, especially when it provides a means of explaining the discrepancy between precession at 25,800yr and the well known precession-related climate cycle with 22,000yr period. Attempts have been made by other theorists in the recent past to attribute this discrepancy to a rotation of the Earth's orbit ellipse with respect to precession, but the results have been quite unsatisfactory. It could be argued that it is not important. It is not something that is going to affect our daily lives, so who cares?

It is important of course because it helps to explain the glaciation data and that provides greater understanding of how climate change occurred in the past and how it may change in the future. The report below is as published in Theoretical and Applied Climatology.

# 3. A 30,000yr Precession-Related Cycle Affecting Climate

## L. G. Bell

*Abstract*

*Evidence is presented for a previously unknown climate cycle of 30,000yr period. The cycle is deemed to be related to the gyroscopic precession, or wobble of the Earth axis. Since it inhibits glaciation, the 30,000yr cycle is called the summer cycle while its counterpart, the 22,000yr "precession" cycle, is the winter cycle. Because of the aspect presented to the Sun, summer is effectively longer than winter. This is used to explain the difference between summer and winter cycle periods and that of the wobble. Some of the problems encountered in interpreting oxygen ratio glaciation data are resolved by knowledge of the existence of the summer cycle and a means is devised for determining the relative volume of glacial ice. An argument is made that essentially eliminates the 100,000yr-orbit cycle, by itself or in combination with other attitude cycles, as a possible cause of the glacial/interglacial cycles.*

## 1 INTRODUCTION

This is the third in a series of reports on long term climate change. It has been undertaken mainly because of problems encountered in data interpretation in the first two reports (Bell, 2002a,b).

The Earth's attitude in space with respect to the Sun has long been thought to have a major effect on climate. In particular a 22,000yr cycle, commonly referred to as the "precession" cycle, with climate effects calculated by Milankovic, is the one most commonly mentioned. In Bell (2002a) the 22,000yr cycle was found to harmonize with the ocean/atmosphere temperature cycle, and thus cause periods of enhanced glaciation.

The purpose of the present report is to present evidence for the existence of a precession-related 30,000yr cycle, to propose a possible explanation of how the precession climate cycles occur and to investigate the implications.

## 2 EVIDENCE

Glaciation data for the last 800,000yr reported by Porter (1989) and published in Encyclopaedia Britannica (1997) are shown here in figure 1. The extent of glaciation was determined by measurement of

37

the oxygen isotope ratio in ocean sediment cores. Along with generally increasing glaciation over the 800,000yr-time period, there are many instances of what appears to be sudden deglaciation, 30,000yr apart, and in particular pronounced instances of the effect during each of the last two glacial stages, 120,000yr apart.

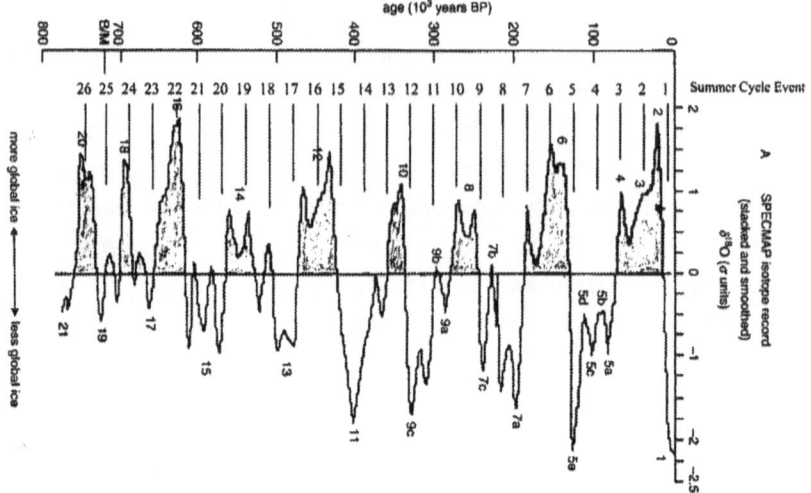

*Figure 1. Marine oxygen isotope records obtained by Porter (1989), as reported in Encyclopaedia Britannica (1997). Positive $^{18}O$ to $^{16}O$ ratio readings in ocean sediments produce shaded areas interpreted as glacial stages. Lines have been drawn and numbered at 30,000yr intervals for the purpose of relating to glaciation rate changes that occur.*

Numbered vertical lines have been drawn at 30,000yr intervals on figure 1 starting with the two particular instances mentioned above and going backward and forward over the remainder of the 800,000yr history. The fourth and fifth glacial stages before present seem to have shifted in time by about 15,000yr, possibly because of time-measurement error. Making adjustments for this presumed error allows apparent deglaciation events to occur during glacial stages at almost every 30,000yr interval. Some appear to end a glacial stage, some to cause glacial retreat, as in the early part of the last two glacial periods and some to have little effect because they happen to occur during an interglacial stage or at maximum glaciation. This provides evidence for a 30,000yr-period heat-cycle of modest strength.

38

## 3 EXPLANATION OF PRECESSION CYCLES

The extension of the axis of the Earth toward the north describes an anti-clockwise circle in the sky with the Pole Star as one point on the circumference. Approximately 26,000yr are required for one complete revolution. This wobble or gyroscopic precession results from a torque supplied by the gravitational forces of the Sun and Moon acting on the slight equatorial bulge (Fowles, 1977). The Earth acts much like a top, or gyroscope, precessing in a stable manner as a result of the constant torque supplied by gravity.

The angular diameter of the cone described by the Earth axis is about 47°, half of which, 23°26' is the obliquity, or angle between the ecliptic, the plane of the Earth's orbit around the Sun, and the plane of the Earth's equator. Thus as the axis precesses around the circle in the sky, the equinoxes and solstices migrate around the Sun, also making one revolution in 26,000yr. Looking down from the stars, summer would be seen to revolve clockwise from one side of the Sun around to the other side and back again. By the laws of mechanics there is no reason to expect the arc transcribed in the sky to be anything other than a perfect circle and therefore no reason for summer to change as it moves around. The obliquity, however is reported (Illingworth et al, 2000) to change between the angles of about 22.1° and 24.5° apparently in step with the precession cycle. (No reason could be found in astronomy texts for precession to be exactly in phase with the precession component of the change in obliquity. The change in obliquity also includes a minor nutation component with a 41,000yr period caused by the gravitational effects of other planets, mainly Venus and Jupiter (Schneider, 1996).)

Comments in astronomy texts suggest that the 22,000yr-"precession" climate cycle is produced by the precession-related change in obliquity. It always has been a problem accepting that a 22,000yr-climate cycle can be related to a 26,000yr-orbital change. Joseph Adhemar published a book in 1842 in which he proposed that longer winters alternate between the northern and southern hemispheres every 26,000yr, and because the Earth-orbit ellipse is rotating, the true period is "21,000yr" (Ekman, 1993). The strange thing is that Adhemar, that long ago, recognized this climate change cycle as a winter cycle; however it is hard to imagine how his proposed mechanism could work. The knowledge that there is an additional 30,000yr cycle may allow another explanation. The difficulty, as Adhemar found, is to conserve the laws of time.

The 30,000yr cycle is referred to as the summer cycle because it inhibits glaciation, whereas the 22,000yr cycle is called the winter cycle because it contributes to enhanced glaciation (Bell, 2002a). It is traditional to think of the yearly climate change in terms of four seasons, however, for our purpose we will consider the year divided, unevenly as it turns out, into only two seasons, summer and winter. It is hereby postulated that it is the aspect of the Earth presented to the Sun that is important. More of the hemisphere, whether north or south, is exposed to the Sun in summer than in winter. Summers therefore are stronger and last longer than winters.

Imagine once again looking down at the Earth orbit from the north at a small angle to the ecliptic; the Sun in the middle with the Earth revolving around anti-clockwise. The seasons precess around clockwise doing a circuit in 26,000yr. As the northern summer solstice approaches the extreme at your right hand, the Earth tilts more than normal toward the Sun. It comes toward you, passes the midpoint then proceeds toward the left side where the Earth axis is by this time tilting a bit farther than normal away from the Sun. This describes the 26,000yr change in obliquity.

As summer proceeds on the back side, because it is the stronger of the two seasons, summers of individual Earth revolutions begin to show a heating trend 1000yr before the mid point. When the season reaches the right hand side, summer heating peaks then coming toward you it again passes the midpoint; summer heating effect is diminished but carries on for another thousand years. The axis tilt at the right has thus caused a positive heating cycle up to a maximum and back in 15,000yr. As the Earth continues to revolve summers start to notice a cooling trend 1000yr before the midpoint on the near side of the circuit. Cooling reaches a maximum at the left and peters out 1000yr past the mid-point on the back. The axis tilt at the left has thus caused a negative summer heating cycle, down to a minimum and back over a second 15,000yr. Together the two halves make up the 30,000yr summer-climate cycle.

While this is happening, southern winter begins a cooling trend 1000yr past the back mid point, proceeding down to a minimum and back in 11,000yr to produce the first half cycle. A heating trend then starts 1,000yr after the forward mid-point, completing the second half cycle by going to a maximum then petering out 1000yr before the back mid point. Together the two halves make up the 22,000yr-winter climate cycle.

In each case as soon as one cycle is over, the next one starts causing a continuing summer cycle every 30,000yr and a winter cycle every 22,000yr.

Southern Hemisphere cycles occur at the same time as Northern Hemisphere cycles since the change in obliquity is the same for both. This being so we may take advantage of the fact that summer in the Northern Hemisphere pays off against winter in the Southern Hemisphere and assume that summer cycles also pay off against winter cycles. The exact mechanism remains unclear, however while northern summer is undergoing the heat cycle of 15,000yr starting *before* the rear mid point, southern winter starts an 11,000yr cooling cycle *after* the rear midpoint. The time-discrepancy for one cycle is canceled by that of the other and the laws of time are thus preserved.

## 4 INTERACTION OF ATTITUDE CYCLES

The Earth's orbit changes from a circle to an ellipse and back to a circle in about 100,000yr. When it is an ellipse it is still close to a circle but with sufficient difference to allow about 7% more solar heating at closest approach, perihelion, than at aphelion. It has been claimed for some time that interaction with other attitude cycles makes use of this difference to produce a 100,000yr-heat cycle and this is what determines the timing of glacial periods.

As mentioned above Southern Hemisphere cycles occur at the same time as Northern Hemisphere cycles. However as far as orbit interactions are concerned the two are exactly opposite. If northern summer for example occurs at the perihelion of the orbit cycle, southern summer is at the aphelion. All things being equal the two should effectively cancel each other. But they are not equal. The Northern Hemisphere has the greater landmass and hence responds dramatically to changes in heat input. The vast expanse of ocean in the Southern Hemisphere, on the other hand, moderates the climate and prevents the same theoretical extremes from occurring (Collier's Encyclopedia, 1996). The north is where most of major glaciation events occur and no doubt it is the northern summer cycle that is responsible for the apparent deglaciation events witnessed by oxygen ratio measurements. Since they dominate, climates referred to herein are those of the Northern Hemisphere unless otherwise stated.

There is such a thing as a season being in-phase with the orbit cycle. If, for example, the maximum summer climate effect due to precession should happen to occur at or near perihelion of the ellipse stage of the orbit cycle, the season would be in-phase. Heating effects of the two attitude cycles would add together. Summer would subsequently precess around to the aphelion where the two cycles would again add to produce a double minimum. By the time summer returned to the perihelion, the orbit cycle would be well on its way to the circle stage but the two attitude cycles might still act in concert to produce a larger than normal heating effect.

With the orbit cycle being 100,000yr and the precession cycle 26,000yr, or more accurately 25,800yr (Bishop, 2001), a season in-phase with perihelion for one maximum ellipse stage, would in four precession circuits, be back to perihelion but 3200yr out of phase. After two orbit-cycles the season would be out of phase by 6400yr, which would mean that by this time it would be the mid-point of seasonal precession that was in phase. It is unlikely therefore that a season could be considered to be in phase for more than two orbit cycles at the most. If the approximate 100,000yr-orbit cycle were to be actually 103,200yr, the season would remain in-phase in every successive orbit cycle. For argument's sake, as unlikely as it is, let us assume that seasons do remain in phase thus allowing similar things to happen every orbit cycle.

The only way to have a climate change as a result of the orbit cycle is to have one of the precession related extreme seasons, summer or winter, occurring at or near the perihelion when the orbit is in the ellipse stage. If it occurs at any other point, circle or ellipse, there is little effect. A stylized depiction of how both the summer and winter cycles might change under the prescribed optimum conditions is shown in figure 2. In this figure the dashed lines represent the maximum and minimum potential contribution effected by the orbit cycle, the upper line being that of a season in phase with the perihelion and the lower line that of a season in phase with the aphelion. As the seasons precess around, there is no orbit contribution to climate cycles when the orbit is a circle. As the orbit changes to an ellipse, seasons in phase with the perihelion as in figure 2Aand 2C, act in concert to produce more temperature fluctuation, the average over time remaining about the same. When the season is in phase with the aphelion as in figure 2B, the attitude cycles oppose each other and there is less fluctuation.

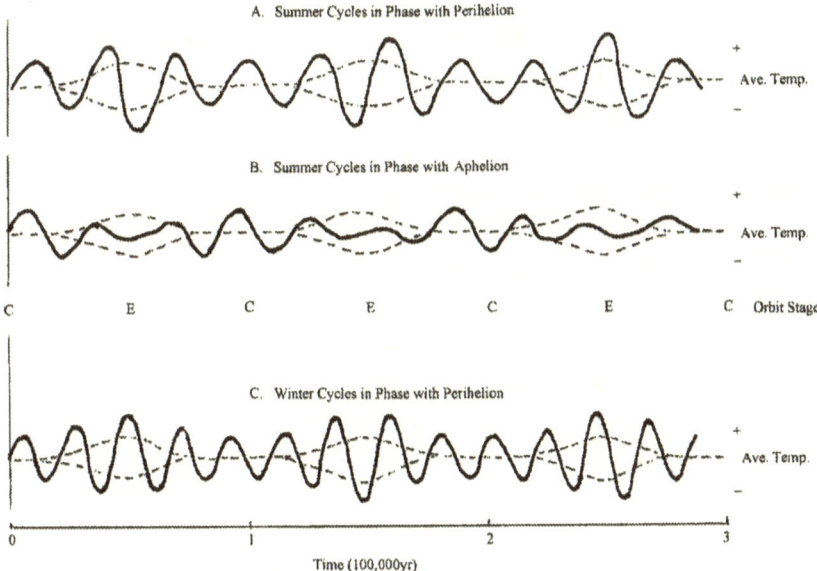

*Figure 2. Stylized depiction of summer and winter temperature cycles as modified by the Earth orbit cycle. The potential orbit-cycle contribution is shown as a dashed line and all contributors are assumed to have the same amplitude.*

The summer cycles could not increase the amount of glaciation. However from arguments made in Bell (2002a) there would be a slightly greater tendency for glaciation at the ellipse stage with winter in phase with the perihelion (Figure 2C) and this would be the only time and the only reason for any orbit-related glaciation effect. The buildup of extra ice during a cold part of a winter cycle, augmented by the orbit cycle, would reflect heat away during the subsequent warm part thus reducing the average temperature. It may thus be concluded that there are two requirements to have the orbit cycle involved in the timing of major glaciation: 1. Northern Hemisphere winter had to have been in phase with the perihelion for at least the last five orbit cycles and 2. The ellipse stage had to, and still has to, occur near the beginning of glacial periods. – The first is unlikely as argued above. As for the second, it is reported (Collier's Encyclopedia, 1996) that we are now near the extreme ellipse phase (perihelion currently falls on January 2 and since the seasons turn clockwise the winter solstice will be exactly at perihelion in about 800yr). This means that Northern Hemisphere winter is in phase with the perihelion, however this is not the start of a glacial stage and therefore the second condition is not being met. Cooperation of the attitude cycles cannot therefore be the cause of the glacial/interglacial cycles. As argued in Bell (2002b) the timing of glacial stages is determined by the ocean temperature cycle. Glacial periods in recent times just happen to roughly coincide with one of the stages of the orbit cycle.

The theory of climate-control oscillation espoused in this series of reports requires that cycles generally grow with time. Power spectrum evidence is presented by Peltier (2001) that during the last million years glaciation events occurred at regular intervals of 100,000yr. This would seem to contradict the oscillation theory, and therefore perhaps support the orbit theory. However there was no indication in Peltier's analysis of the 100,000yr-power spectrum during the previous million years. It appears therefore that climate cycles did indeed grow according to the oscillation theory, up to a point, then sometime after the beginning of the latest million years the cycles stopped increasing, at least in period duration. There are several reasons why this should have happened:

1. As theorized in Bell (2002a) a grand harmonic node occurred at about 700,000yr BP and since then the degree of glaciation has been decreasing thus offsetting the expected gradual increase and tending to make cycles the same.
2. An harmonic node of 352,000yr period, as described in Bell (2002a), happened during the fifth glacial stage

before present thus making that stage larger than it would otherwise have been. The two subsequent stages were smaller, not being under nodal influence. Another 352,000yr node occurred during the interglacial period between the last two glacial stages making these two stages intermediate in size and also about the same in most other respects. The influence of harmonic nodes is apparent in a set of the longer-term glaciation data of Porter (1998) shown here in figure 3.

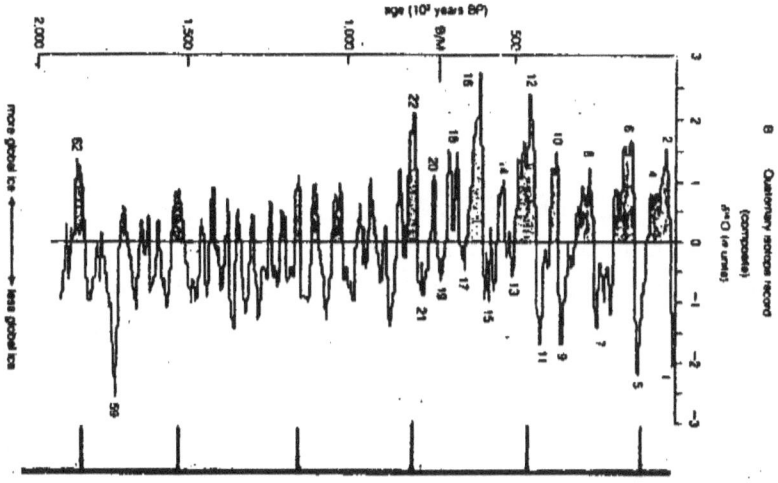

Harmonic glaciation nodes at about 352,000 yr intervals.

*Figure 3. Longer-term record of glaciation over previous 2myr as measured by Porter (1989) using the ratio of $^{18}O$ to $^{16}O$ in ocean sediments and showing harmonic nodes at marked intervals.*

3.  As theorized in Bell (2002b), regrowth of the northern biomass at the beginning of an interglacial period stops the increase of atmospheric $CO_2$ concentration thus limiting the temperature increase. Biomass intervention would always come into play at approximately the same climate conditions and would thus produce about the same temperature profile in successive interglacial periods.

All of these factors would tend to influence glacial cycles and possibly limit period length. It is just a coincidence that this limit is 100,000yr, which has led us to wrongly assume that the orbit cycle controls the glacial/interglacial cycle period.

## 5 ICE VOLUME

There has been reluctance to refer to the change in glaciation associated with the summer cycle as deglaciation. To cause melting near the beginning of the Ice Age for example, heat from the summer cycle would have to be sufficient to overcome albedo cooling by the ice as well as the very cold climate and the cloudy conditions thought to prevail. This would seem to be almost impossible for a heat cycle found by other indications to be relatively weak. Consideration of this problem led to the realization that the oxygen-ratio data produces a measure of the **rate** of glaciation rather than the **amount**. A slice of ice core would represent the increment of ice added to the glacier during the interval of time represented by the slice, not the total amount of ice accumulated up to that point. The oxygen-ratio measurement is therefore a rate of growth and comparison with subsequent measurements produces a rate change, an acceleration or deceleration of growth. Ice accumulation continues as long as the ratio is positive and reduction in total ice-volume occurs only when the ratio is negative.

Knowing that the oxygen-ratio data is a rate allows for a rough calculation of the relative amount of ice for various stages of the glacial cycle. Rate can be changed to amount by integration. To do this the oxygen-ratio data for the last two glacial and interglacial periods from figure 1 were magnified again by a factor of 4. Vertical lines were drawn on the expanded graph and measurements taken so as to be able to reduce the Y-axis and expand the X-axis. A sample result of this procedure for the Ice Age-glacial period and the current interglacial period is shown in figure 4.

## Summer Cycle

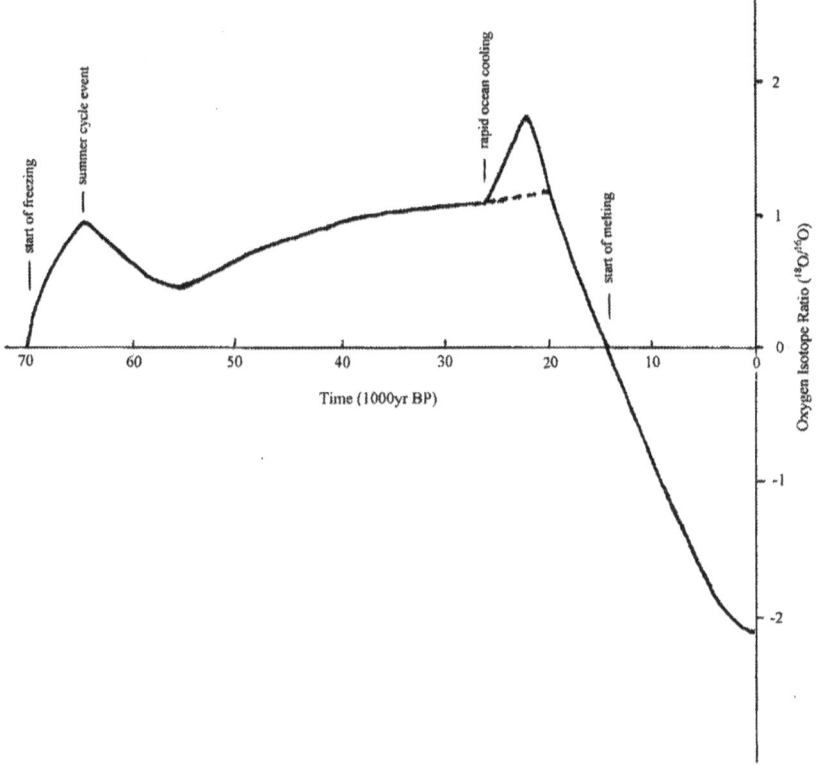

*Figure 4. Oxygen isotope data for the Ice Age and the present interglacial stage. The data of Porter (1989) from figure 1 for this period of time has been reduced in the Y direction and expanded in the X direction.*

The modified oxygen-ratio data, regarded as being the first derivative of glacier volume, was integrated to obtain the amount profile shown in figure 5. This was done by dividing the rate graphs produced, as in figure 4, into time segments, estimating the average rate of each slice, multiplying by the slice thickness to obtain an amount for the slice and adding progressively. The second peak in the oxygen-ratio curve for the Ice Age was not included in the calculation, the dashed line being followed instead. The reason for this, is that this peak is believed to be the result of ocean cooling rather than increased glaciation (Bell, 2002b). The starting amount of ice volume was taken to be zero but obviously there is a base amount in the Polar Regions, that which provides ice-core data, that never melts.

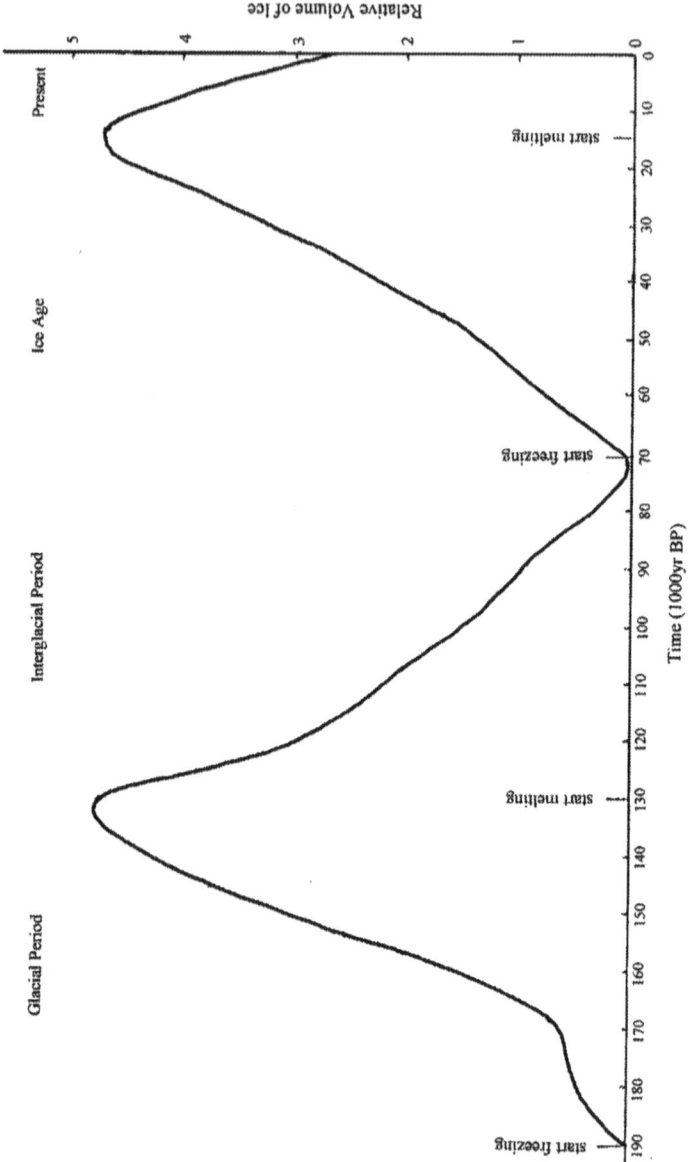

*Figure 5. Relative volume of glacial ice above base amount for last two glacial and interglacial periods as obtained by integration of oxygen isotope data, the latter being taken to be a rate measurement.*

Summer cycle events are noticeable on the ice-volume profile in figure 5 but only just for the Ice Age-glacial stage. If the oxygen-ratio data were not a rate measurement the summer cycle would have been missed.

It is surprising to note that the amount of ice is about the same for the interglacial period as for the glacial period. On the other hand ice that freezes has to thaw and since times are about the same the average amount has to be about the same. Albedo cooling of the ice is overcome in the interglacial period by clear skies to start, by increased $CO_2$ and by a lagging ocean-temperature cycle. The resulting warmer climate allows the ice to melt. All of this is referred to in Bell (2002b). Nothing new is added to the present general theory of climate change by these ice-volume profiles; however they are interesting in themselves. They dispel, for example, any suggestion that major retreats and re-advances occur during growth stages. These must occur, as mentioned before (Bell, 2002b), as a result of plastic flow of the ice after melting has started. The data also suggest that we should currently have about 57% of the maximum glacial-stage ice.

## 6 GLACIATION EFFECTS

One of the things that could not be explained in Bell (2002b) was the apparent deglaciation occurring in the middle of the Ice Age. It was the inability to explain this particular observation that led to the discovery of the summer cycle. The apparent retreat has now been explained. It is one of the summer cycles and a retreat does not actually occur, rather it is merely a change in rate of glaciation as a result of extra heat from the summer cycle. Figure 1 of the current report shows that the second summer cycle during the Ice Age did not produce as dramatic a change in the rate of glaciation, as did the first one. It would seem to have been too cold owing to the fact that the amount of ice had increased by a factor of about 5 during the intervening 30,000yr (Figure 5). A summer cycle was seen on the other hand, for reasons unknown, to cause a change in rate at the maximum of the previous glacial stage. Perhaps in this latter instance it was the temperature-related [18]O selection by microorganisms that was affected by the summer cycle.

The theory described in previous reports in this series seemed to demand a gradual and orderly increase in both period and amplitude of glacial/interglacial periods. The record by oxygen-ratio measurements, however, showed very wide scatter. The reason for this is also now clear. With the technique used, glacial stages appear to be chopped in two or three parts and sometimes ended by the heat effect of the summer cycles but this is not what actually happened. Summer cycles only cause a slowdown in glaciation and oxygen-isotope-ratio measurements seriously exaggerate their effect. This would tend to produce the scatter that is observed.

## 7 PRESENT TIME

The last clear indication of a summer cycle event is about 64,000yr before present (BP). That would mean that an event came to pass 60,000yr later at about 4,000yr BP. It is hard to say when, during the cycle, the event occurs. It seems most likely that it happens near the maximum of the summer cycle but it could be several thousand years earlier and even though it is occurring at the aphelion of the orbit cycle, the climate could still be affected by it.

An attempt was made to determine the relative contribution of orbit and precession cycles by analyzing the effect of the summer cycle on glaciation. Taking the 100,000yr-orbit cycle at face value, as argued above, seasons would go completely out of phase, changing from summer at aphelion, as it is now, to summer at perihelion in 4 orbit cycles. This should make it possible to determine relative contributions. The problem with this plan is that the summer cycle effect over the 700,000yr-period, for which there is data, is observed to be fairly constant. This is because the ellipse stage of the orbit cycle has been occurring during interglacial periods; thus the only data available is that for the circle stage.

## 8 CONCLUSIONS

1. There is a summer climate cycle of 30,000yr period.
2. It is postulated that this summer cycle, like the previously known 22,000yr "precession" or winter cycle is related to the 26,000yr wobble of the Earth axis, the period of the latter being modified by the relative duration of the two seasons.
3. Neither the 100,000yr-orbit cycle nor any combination with other attitude cycles is likely to have been the cause of the glacial/interglacial cycles.
4. The period and amplitude of glaciation cycles appear to have reached a limit possibly because of harmonic

node effects or because biomass intervention inhibits further growth.

5. Oxygen ratio data are found to indirectly measure the rate of ice accumulation rather than the amount.

6. The effect of the summer cycle is exaggerated in oxygen ratio data. This solves problems with data interpretation in previous reports.

7. One of the summer cycle deglaciation events was scheduled to occur about 4,000yr ago and could be having an effect on current climate.

8. The oxygen ratio data, as a rate measurement, may be integrated to obtain an ice-volume profile for the glaciation cycles.

Acknowledgments

The author gratefully acknowledges kind permission by *Quaternary Research* to republish data in figure 1. Advice and support from J. C. Anderson and editorial comment from J. A. L. Robertson is also appreciated.

References:

Bell LG (2002a) World Ocean Temperature Lag Time, An Analysis Based on Glaciation Data for the Last Two Million Years. (to be published in the Journal of Theoretical and Applied Climatology).

Bell LG (2002b) Ice Age Mystery, A Proposed Theory for the Cause of Long Term Climate Change. (in press).

Bishop R (2001) Some astronomical and Physical Data, The Observers Handbook of the Royal Astronomical Society of Canada. R Gupta ed. Toronto: Toronto University Press: 25.

Collier's Encyclopedia 1996, 20: 561-563.

Encyclopaedia Britannica 1997, 19: 860-868.

Ekman M (1993) A Concise History of the Theories of Tides, Precession, Nutation and Polar Motion (From Antiquity to 1950). Surveys in Geophysics 14: 603,605.

Fowles GR (1977) Analytical Mechanics, Third Edition. New York: Holt, Rinehart and Winston, Ch. 8: 243.

Illingworth V, Clark JOE (2000) Facts on File, Dictionary of Astronomy. New York: Checkmark Books: p293.

Peltier WR (2001) Earth System History, Encyclopedia of Global Environmental Change, Munn RE ed. Chichester: John Wiley and Sons, 1: 45-46.

Porter SC (1989) Some Geological Implications of Average Quaternary Glacial Conditions. Quaternary Research. 32: 245-261, Figure 1: b and c

Schneider SH ed (1996) Encyclopedia of Climate Change and Weather, Oxford University Press, Vol. 2, Milankovitch: 507.

The explanation for summer and winter cycles, as described in this Summer Cycle report is not yet completely satisfactory to this author. It is easy to believe that in any particular precession circuit there could be a summer cycle lasting 30,000yr and a winter cycle lasting 22,000yr, the two averaging out to 26,000yr, (an approximation of 25,800yr). The next precession circuit the same thing happens. But if you then try to plot them both on the same time line you have more winter cycles than summer and they are not pared. It has to happen that way but it is difficult to understand and accept.

One perhaps has to accept time-sharing between summer and winter cycles and between northern and southern hemispheres. With this, one can imagine that it is possible to satisfy the laws of time. This is not exactly a new idea. Joseph Adhemar almost got it right in 1842 when he suggested that his precession-related cycle, thought at the time to have a period of 21,000yr, is a winter cycle and flip-flops from north to south. Unfortunately he didn't have evidence for a 30,000yr-summer cycle and couldn't complete the argument.

# Chapter 4
# Temperature Disparity Effects

In the Ice Age report of Chapter 2, it was found, among many other things, that at the start of an interglacial period, $CO_2$ is released from the ocean by a change from skies that were relatively cloudy during the glacial stage to skies that are relatively clear. It was taken for granted that $CO_2$ was released from the ocean by the heat from extra sunlight. This is the generally accepted view in mainstream climatology, however no explanation is available to explain why the buildup should come to an end, and also how and why $CO_2$ is eventually reabsorbed into the ocean. After studying the curves for many months it became apparent that it was the temperature disparity between atmosphere and ocean that determines the rate and direction of $CO_2$ change. The temperature disparity report pasted below is the final result after many revisions.

As mentioned above when writing the Ice Age report it was accepted that heat causes release of $CO_2$. This turned out to be wrong. It is demonstrated in later reports that it is not heat that causes release, but cold.

This fits the data much better. The Ice Age report and two others were published in Theoretical and Applied Climatology and once published the words are cut in stone. To change them now would be changing history and would also require rewriting subsequent reports. To leave them unchanged also provides some sense of how the ideas evolved.

It was found quite impossible to get the present report published in a science journal. It was approved for TAC but I found what I considered to be a fatal error and withdrew it. When I resubmitted, having corrected the error, I found that clientele had changed and it was rejected. It has since then been rejected by many others. The main problem with the reviewers is the suggestion that $CO_2$ is released from the ocean by a thermal diffusion effect. They invariably say that molecules of a compound like $CO_2$ can *only be induced to move by a concentration gradient of that compound, not something else, like temperature difference*. It is obvious to me however that a temperature gradient, or temperature difference across a semi-permeable barrier or even a pressure gradient (or a concentration gradient for that matter, as with osmosis) can cause a migration, how otherwise could one explain condensation. In this case water molecules in air migrate across the interface from a relatively low concentration to 100% $H_2O$ only because of temperature difference. $CO_2$ in water or air is a good candidate for this mechanism because it is a relatively large molecule. From Brownian movement we know that molecules are bouncing around in a gas or liquid, and even to some extent in solids. It seems obvious that larger molecules moving inadvertently into lower temperature or higher pressure are not as likely to move back. This produces a net flow, a thermal or pressure diffusion effect. I wrote a proposal for someone to hopefully undertake a research program to prove this scientifically and paste a copy to the end of this chapter.

The report of this chapter provides evidence to support the theory that is being developed. The temperature difference between atmosphere and ocean determines the amount of cloud cover and that in turn determines whether there is release or absorption and the rate at which change occurs. The formula for $CO_2$ exchange will prove useful in calculating future conditions.

## 4. Ocean/Atmosphere Temperature Disparity Effects:
An Analysis of $CO_2$ and Temperature Data
for the Previous Interglacial Period

## L. G. Bell

*Abstract*

*Published information is used to confirm that during glacial cycles the release and absorption of $CO_2$ by the ocean can be related to the disparity between land/atmosphere and ocean temperatures. At the beginning of an interglacial period skies become clearer because of ocean temperature lag, thus increasing temperature disparity and creating a feedback, which causes an exponential increase in atmospheric $CO_2$ concentration and temperature. An equation is determined for the rate of $CO_2$ release during this time. Continuing escalation of $CO_2$ concentration is slowed to a linear increase with time and is eventually stopped by biomass intervention thus causing a delay and allowing the ocean temperature to catch up and surpass that of the atmosphere. Skies then become relatively cloudy and $CO_2$ is reduced. It is theorized but not proven that temperature disparity, whether positive or negative, is as effective in changing the amount of $CO_2$ in the biosphere. All effects observed are related to diffusion of molecular $CO_2$ across the air/water interface and to mixing and a proposed thermal or pressure related migration in the ocean.*

## 1 INTRODUCTION

This is one of a series of reports describing a new theory on the cause of climate change. Advantage has been taken of major drastic climate changes of the Pleistocene ice age of the last 2myr, and in particular those of the last major transition back to about 160,000yr BP. Glaciation data from ocean sediment cores (Porter, 1987) have been analyzed as well as $CO_2$ and temperature data from Antarctic ice cores (Barnola et al, 1991, Jouzel et al, 1987 and others (see Boden et al, 1994)). Among many effects it is suggested (Bell, 2003) that major glaciation cycles are caused in part by the release and absorption of $CO_2$ by the ocean. This idea goes against a currently accepted view (e.g. Jacob, 1997) that the biomass acts as the major source and sink and the new theory has therefore met with some skepticism. The idea has however been given further strength by recent analysis of $CO_2$

measurements at remote sites (Bell, 2005). It is found that the ocean must indeed be regarded as the major source and sink and exchange between the two media is the result of diffusion across the air/water interface.

For word economy the "rest of the environment", as opposed to the ocean, may be referred to herein as the "atmosphere". It is found in earlier reports that the ocean temperature lags behind that of the rest of the environment/atmosphere by about 12,000yr (Bell, 2002 & 2003). It is postulated that when the ocean is relatively cool the climate is clear and when it is relatively warm the climate is cloudy. The result of this is that during a glacial stage, with the ocean lagging behind as the atmosphere cools, the climate is moderately cloudy. With air temperatures continually falling, as a consequence of the cloudy conditions as well as feedback cooling from glaciation, the ocean eventually begins to freeze. This causes the ocean temperature to fall below that of the atmosphere and skies start to become clear. The atmosphere starts to warm up, the ice begins to melt and the glacial stage comes to an end.

The ocean continues to cool after the air/sea temperature crossover at the start of the interglacial stage because of glacial runoff. Skies continue to clear; there being less evaporation from a cooler ocean and less condensation to form clouds in a warmer atmosphere. In this circumstance, when skies are relatively clear, $CO_2$ comes out of the ocean. It was assumed in Bell, (2003) that clearer skies allow the Sun to heat the surface of the ocean and thus release more $CO_2$ where the Sun is shining, than is being absorbed where it is not, thus producing a buildup in the atmosphere. Recent work mentioned above (Bell, 2005) has shown that it is the air temperature, rather than the amount of sunlight, that determines the $CO_2$ flux across the interface and that clear skies are effective in causing buildup by reducing winter temperatures in the far north. Only after about 20,000yr of a normal interglacial period does the ocean warm enough so as to produce extra cloud cover, and thus allow more $CO_2$ to be absorbed than is released.

After a temperature crossover there is positive feedback. When the atmosphere becomes warmer than the ocean for example, change is from cloudy to clear; $CO_2$ is released, which warms the atmosphere thus producing clearer skies and consequently more release. As the temperature difference increases so does the rate. Another way to look at it is that after a climate change, ocean/atmosphere temperature disparity increases and this is what determines the rates of release and absorption.

The main objectives of the present work are to examine more carefully the release and absorption of $CO_2$ from and by the ocean as related to temperature disparity, and if possible to determine an equation for change.

## 2 INFORMATION

It is espoused in this series of reports that, over the greater part of the Earth's history, climate control has been achieved by the amount of cloud cover and that the control system went into oscillation to initiate the Pleistocene era. In Bell (2002) harmonic beats observed in the Pleistocene glaciation data of Porter (1987), led to the conclusion that ocean and atmosphere temperatures began to cycle opposite to each other as continents moved to their current location, and have continued to do so right to the present time. The temperature lag time, for temperature cycles associated with the harmonic beats, was accurately determined to be 11,750yr.

Comparative temperature and $CO_2$ concentration profiles obtained from Antarctic ice cores for the early part of the previous interglacial period were published by Barnola et al (1991) and are reproduced here as figure1. As with $CO_2$ data, the temperature profile for Antarctica is taken to represent global change. This may be invalid however it is the only temperature data we have.

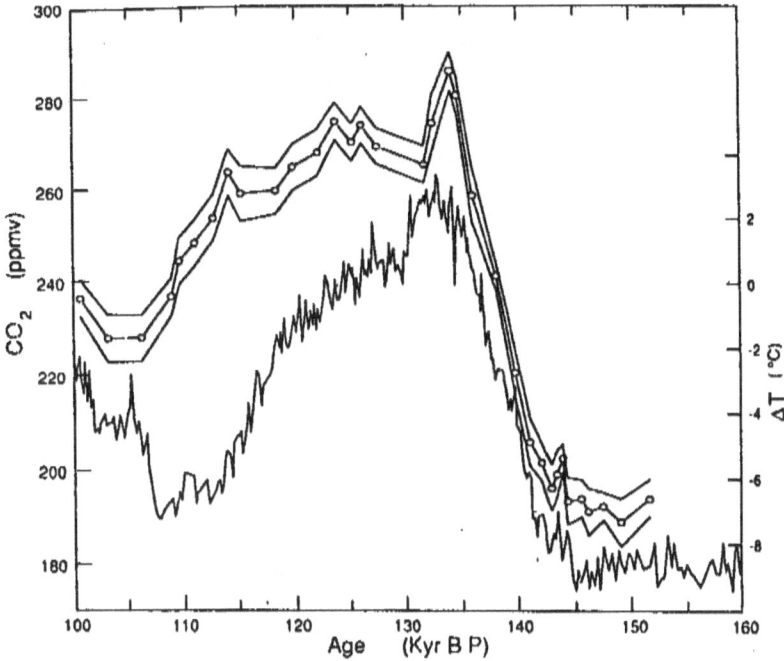

*Figure 1. $CO_2$ and antarctic surface temperature variations recorded on the Vostok ice core and presented by Barnola et al (1991). The $CO_2$ plot also shows the envelope of measurement uncertainty.*

## 3 PROCEDURE AND OBSERVATIONS

### 3.1 Disparity calculations

The time scale in figure 1 was reversed and median lines of temperature and $CO_2$ data were used to produce figure 2. The scale for the temperature graph was raised and expanded by a factor of 1.4, but the data was not shifted in time. Raising the temperature scale allows the two graphs to merge at the start of the period under investigation in the neighborhood of 150kyr BP. The ocean temperature supposedly lags behind that of the atmosphere by approximately 12,000yr as mentioned above. The reason for choosing a factor of 1.4 is that it allows the temperature and $CO_2$ profiles to cross a second time 12,000yr after the increasing temperature curve. Doing this also allows the $CO_2$ curve to define the profile of the global warming effect of this variable assuming a linear relationship between $CO_2$ concentration and temperature. (These changes also negate the requirement in previous work (Bell, 2003) of a lag of global temperature on heating and lead on cooling of 1500yr cf. Antarctic surface). Note that the curves cross a third time, in this instance 12,000yr after the decreasing atmosphere temperature profile. The third junction is not as well defined as the second, however there is considerable scatter in the data at this point.

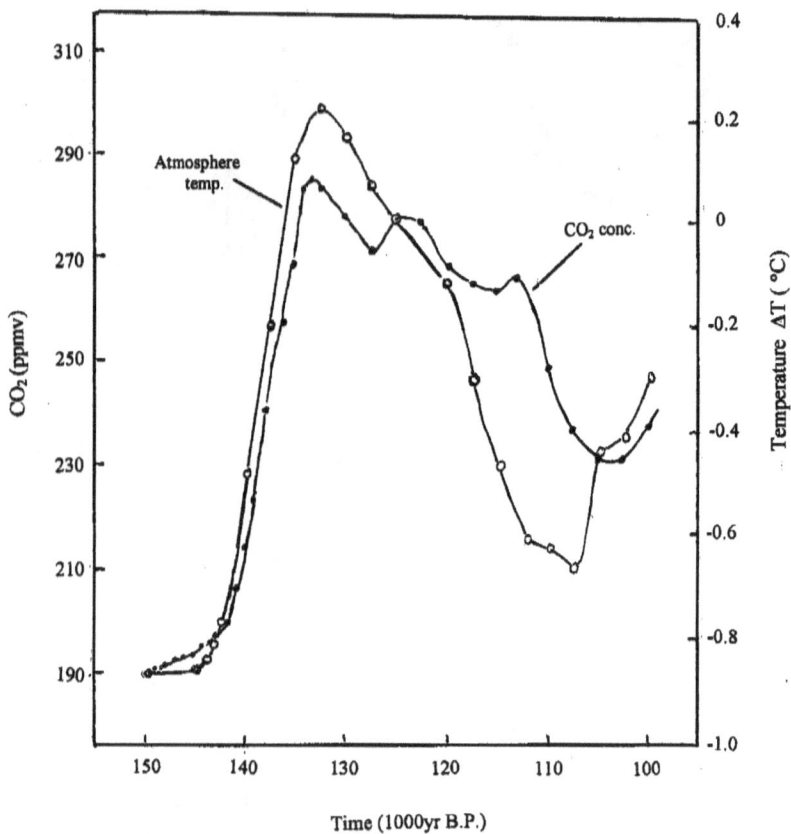

*Figure 2. The time scale was reversed for the Antarctic ice-core data in figure 1. Median $CO_2$ concentration and temperature profiles were replotted with the surface temperature profile raised and the scale increased by 40%.*

The shape of the ocean temperature profile is important to this analysis and is not accurately known. For this exercise the early part of the *ocean* temperature/time profile was taken to be that deduced from changes thought to have been responsible for ending the Ice Age (Bell, 2003). It was then assumed that it would continue to cool for some period of time but to eventually recover and then follow the *atmosphere* temperature with a lag time of 12,000yr but be slightly less volatile (figure 3). Expected climate conditions are indicated. When skies are relatively clear, a cloud cover factor *increases* the temperature above that expected from the $CO_2$ concentration-contribution alone. When the skies become relatively cloudy, the cloud cover factor *reduces* the temperature below that expected from $CO_2$. (It is to be observed in Bell (2005) that the cloud cover factor may be a change in temperature associated with the change in the Earth's albedo). The ocean temperature profile passes through the junction of the other two curves for atmosphere temperature and $CO_2$ concentration at all three locations. This is expected because, according to the theory, $CO_2$ is released or absorbed by temperature disparity between atmosphere and ocean. If there is no disparity, there is no contribution from other factors, thus the temperature is defined by $CO_2$ and the three curves cross together. The fact that they do so on three occasions validates this approach.

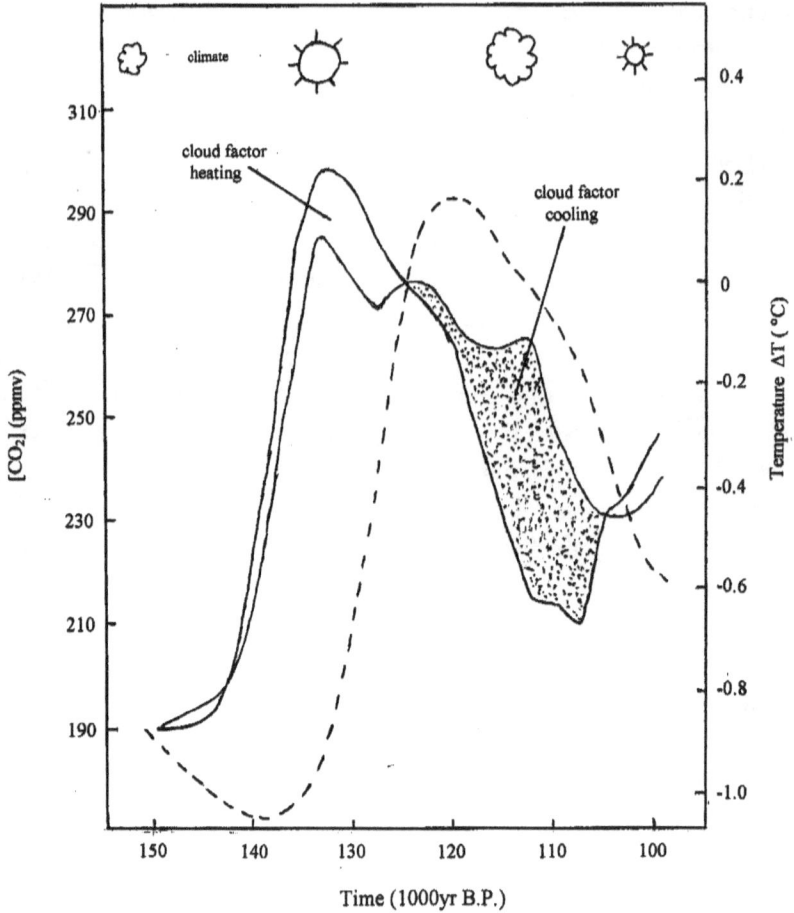

*Figure 3. Atmospheric CO₂ concentration and surface temperature profiles are copied from figure 2. Also included in this figure is the projected ocean temperature with 12,000yr-lag time. Expected climate conditions and cloud cover factor contribution areas are indicated.*

It is realized that release and absorption are rate processes and that integration is required to calculate the concentration profile. To test the disparity theory, the early part of the heat up curve, after 151kyr BP, was divided into intervals of 2000yr. Using $CO_2$ concentration of 190ppmv as a base for the initial temperature crossover, contributions were determined for the average of each interval and added progressively. To calculate a contribution, the average disparity δ between ocean and atmosphere temperature for the interval was multiplied by a constant amount C* for each 1°C disparity. The formula for this is:

$$\Delta[CO_2] \approx \sum C^*\delta \qquad\qquad 1.$$

It was found that 22.6ppmv/2000yr°C for C* produces a very good match with measurements. The resulting calculated profile for the first 10,000yr is shown, alternated with measured points, in figure 4. Note that the atmosphere temperature does not begin to respond to the increasing $CO_2$ for the first 5,000yr, possibly because pack ice is still spreading out on the ocean during this time. The ocean is cooling causing greater disparity and higher $CO_2$ release, but the excess $CO_2$ is ineffective in heating the atmosphere because of the ice.

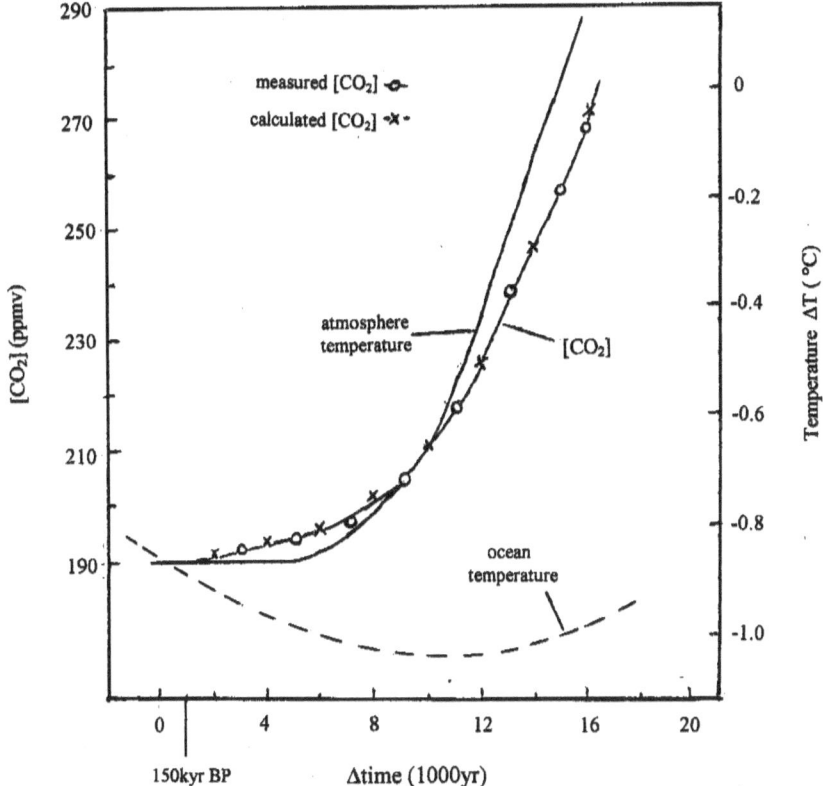

*Figure 4. A close-up view of the early part of the previous interglacial period showing measured $CO_2$ concentration and atmosphere temperature profiles as well as the expected ocean temperature profile. Measured values of $CO_2$ are interspersed with those calculated from temperature disparity.*

Following the success of $[CO_2]$ calculations for the early part of the heat up curve, calculations were continued for the whole period under investigation with results shown in figure 5. It will be noted that the calculated concentration continues to follow the buildup then rises well above the measured amount in the atmosphere, reaching a maximum of 342ppmv at the time of the middle temperature crossover. The amount then drops off rapidly to about 250ppmv at the third crossover.

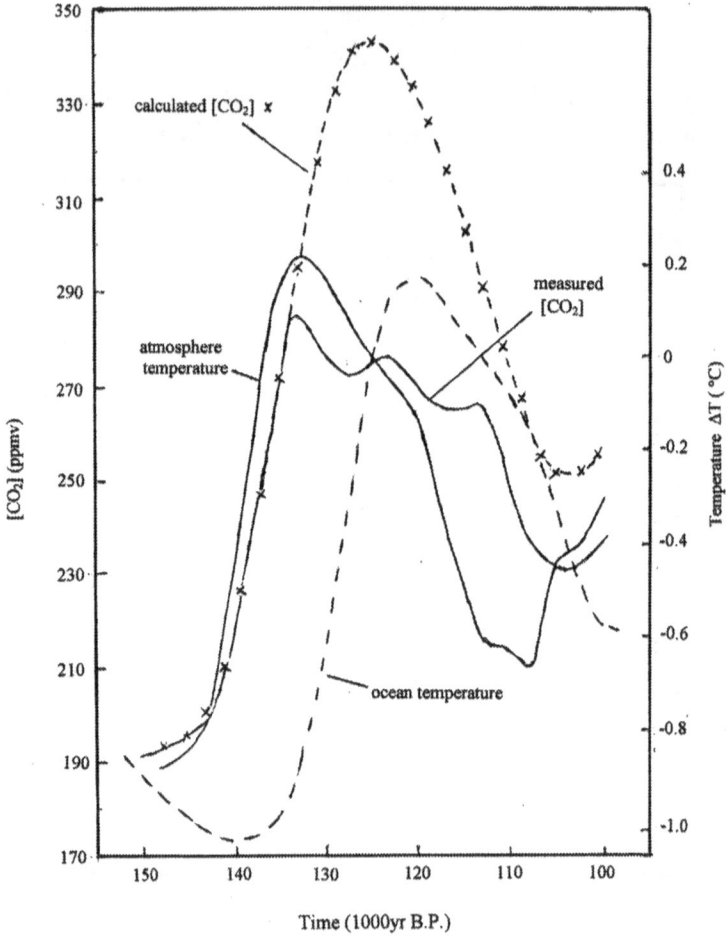

*Figure 5. Temperature and measured $CO_2$ concentration profiles for the start of the last interglacial period copied from figure 3. A calculated $CO_2$ concentration profile has been added. The amount between the measured and calculated curves is thought to be that contained by the displaced northern biomass.*

3.2 Release Formula

To find an equation for $CO_2$ increase, data from figure 1 was carefully re-plotted as a function of time for the early part of the period under consideration as shown in figure 6. The best fit of many that were tried is a function of $t^4$.

$$\Delta[CO_2] \approx 0.0016t^4 \qquad\qquad 2.$$

Differentiating gives a rate of release at time t:

$$d\Delta[CO_2]/dt = 0.0064t^3 \qquad\qquad 3.$$

where t is in units of 1000yr. There is a deviation from formula 2 in the early part of the curve, the increase in $[CO_2]$ being almost linear with time, possibly because the ocean temperature is decreasing, whereas the atmosphere temperature is constant. Above $\Delta t$ of 11 the measured values also increase above $t^4$, indicating a higher exponent of t, then above $\Delta t$ of 12 measured values revert, to a slightly lower rate and a linear relationship with time.

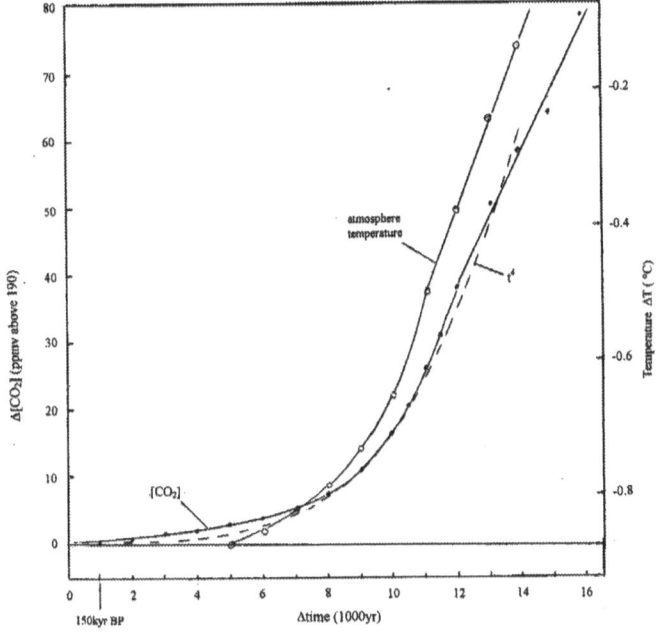

*Figure 6. Measured atmosphere $CO_2$ and temperature at the beginning of the last interglacial period with an attempt to fit the $CO_2$ increase to a $t^4$ function. A deviation from $t^4$ is observed at high and low values of t. The change with time eventually becomes linear.*

# 4 ANALYSIS

## 4.1 Exponential release

Indications are that $CO_2$ buildup at the beginning of an interglacial period is indeed the result of one or more feedback mechanisms. Equation 2 for early buildup having a fourth power term is severe enough, but the measured values seem to increase even above that before the change to a linear relationship. The formula could therefore be considered reasonably conservative for this initial stage of buildup. It has been shown that temperature disparity causes release and that that in turn provides atmosphere temperature increase, from both a greenhouse effect and a loss of albedo. It is not surprising therefore, that release is observed to be an aggressive exponential function of time.

Although the formula may be conservative the increase with time eventually becomes linear as shown in figures 2,3,5 and 6. If $t^4$ were to hold for 16,000yr, $\Delta[CO_2]$ would be about 105ppmv, but with the constant rate of increase of about 10ppmv/1000yr for the linear portion, the measured amount is found to be only about 80ppmv. One would expect exponential release once started to continue and it is of interest to speculate on what would motivate the change.

## 4.2 Change from exponential to linear

Absorption of $CO_2$ by the biomass has to be considered as a possibility. Because of biomass lag, this proposed mechanism has the advantage of not being expected to interfere with the preceding exponential increase. One might also think that it could gradually increase in effectiveness with increasing disparity so as to maintain a controlled linear release. But could it? Biomass intervention is a wild card, not under the control of other variables. Therefore, when it starts to have an effect, it should cause a marked change as it is shown eventually to do.

A second possibility has to do with a concentration effect on interface diffusion. The idea is that $CO_2$ is released from northern oceans and, as concentration builds up, more of it is reabsorbed back into the ocean at median and southern latitudes because of concentration difference. There is little doubt that this happens. It should have a moderating effect on the release rate and be more effective as $CO_2$ concentration increases. This mechanism also has the advantage of being a controlled process, however it should affect release before as well as after the change and therefore cannot be the answer to the question at hand.

A third possibility has to do with regulation of $CO_2$ supply. There is no doubt that the $CO_2$ required to produce the buildup in the atmosphere comes out of the ocean. To meet the requirements of Henry's law of solubility, as the concentration in the atmosphere increases, that of the surface layer of the ocean also has to increase to create a balance across the interface. It is proposed that ocean currents transport $CO_2$ from the ocean depths against gradients of temperature and pressure (see below). It is assumed that there is a certain reservoir in the upper layers that is being supplemented by transport from below and, up to the end of the period of exponential release there is sufficient available to meet the demand. When the release rate becomes too high, however, upper layers become depleted and subsequently the rate is restricted to only that amount that can be transported from the ocean depths. This could result in the change from exponential to somewhat lower linear rate that is observed.

## 4.2 Biomass Intervention

Measured $CO_2$ concentration and that calculated from temperature disparity are in close agreement up to about 18,000yr of the interglacial period. Above that time the calculated value, using equation 1, soars well above the measured value (figure 5). It has to be accepted that biomass intervention is sufficient to stop the increase in atmospheric concentration. The difference between measured and calculated values is the amount temporarily stored in the biomass. $CO_2$ continues to be released from the ocean because of disparity, but is consumed by the increasing northern forest. This moderates the atmosphere temperature, actually causing it to decrease, disparity is reduced and release decreases and eventually stops. It stops at the second ocean/ atmosphere temperature crossover. At crossover there is a balance between absorption and release. As the ocean subsequently continues to warm, thus producing negative disparity, cloudy conditions prevail, release turns into absorption and the total amount of $CO_2$ in the environment (atmosphere and biomass) starts to decline. The 67ppmv that was released and stored in the biomass at the crossover is now partially recovered as the climate cools and forests retreat southwards. As the biomass decays the ocean begins to absorb the resulting $CO_2$. At a certain point the biomass again intervenes, providing more $CO_2$ than can be absorbed, thus warming the atmosphere and allowing the ocean temperature to catch up once again. The calculated total amount of $CO_2$ remains high at the third crossover because some of the northern biomass still survives awaiting further retreat during the ensuing glacial stage.

Incidentally the 67ppmv taken up by the biomass at the middle crossover point allows a calculation to be made of the amount of the northern biomass that is displaced by glaciation. It also allows one to estimate the total biomass of the planet. As a guess the latter might be as much as, but not likely more than ten times the displaced biomass. This would mean that the total biomass is equal to, or less than, that associated with 670ppmv of $CO_2$. (The generally accepted amount of carbon for total biomass is 4000PgC (Jacob, 1999). This works out to an equivalent of about 1800ppmv $CO_2$ should it be introduced to the atmosphere).

4.2 Profile for the last major transition

To check the validity of the assumption that negative disparity is as effective in reducing $CO_2$ in the environment, as positive disparity is in increasing it, one would have to do integration calculations on a complete interglacial/glacial cycle. This was attempted with some success, however the data available is not as accurate for the latter part of the transition, thus there are too many arbitrary adjustments that have to be made to warrant including it here. Analysis would benefit greatly from the use of a computer program since the effect of small changes would then be instantly observed.

5. DISCUSSION

5.1 Disparity effect

It appears from the analysis that temperature disparity is indeed the most important, if not the only factor, determining release and absorption of $CO_2$. It was thought that $CO_2$ additions from volcanic activity, for example, might be noticeable but this does not seem to be the case. We can take from this that release and absorption of $CO_2$ is a diffusion process, an exchange that is effected by temperature difference of air and water in contact. Climate change occurring over thousands of years is the result of a process occurring at the molecular level. As the theory goes, surface tension provides a certain barrier to movement of $CO_2$ molecules across the air/water interface. There is continuous flux in both directions, however warmer molecules are more likely to cross; thus air colder than the water will cause release and provide a pressure to maintain a greater concentration difference. One would think therefore that clearer skies would *increase* the amount of sunlight, warm the atmosphere and cause absorption rather than release. According to the new theory, increased sunlight heats both surface air and water, thus creating a balance for the greater part of the planet. However, in the far north, because of the current particular continental configuration, clear skies disproportionately *reduce* the surface air temperature during winter, thus providing conditions for excessive release and effectively producing a pressure that maintains the resulting buildup (Bell, 2005). When the climate becomes cloudy, extra cloud cover warms the surface air in northern winters thus preventing release and allowing re-absorption. It has not been proven that the negative disparity resulting from cloudy conditions is as effective in changing $CO_2$ concentration as is positive disparity. However to get back to the same situation after each interglacial/glacial cycle as much has to be absorbed as is released because there is nowhere else for the $CO_2$ to go. Choosing the correct ocean temperature profile would provide this result.

5.2 Transport in the ocean

According to Henry's Law of solubility, the amount of $CO_2$ dissolved in water in molecular form depends on the partial pressure of the air in contact and balance across the interface is maintained by diffusion. This means that the $CO_2$ concentration in the water at the surface of the ocean has to vary in proportion to the concentration in the atmosphere. This is probably why only half of the fossil fuel emissions stay in the atmosphere, the other half goes into the upper layers of the ocean to balance that in the atmosphere.

During an interglacial/glacial cycle the $CO_2$ concentration goes from a high of about 280ppmv to a low of 190ppmv. This is a lot of $CO_2$ and the upper layers must have their concentration automatically adjusted to match. Judging by the ratio of 1to1 for fossil fuel emissions, this means that a total of 2x90=180ppmv has to be transported to and away from the surface during each cycle for the atmosphere concentration change. In addition, as calculated from temperature disparity (figure 5), another 67ppmv are absorbed by the biomass and must also be transported to maintain the balance. This latter amount does not have to be matched because it is not balanced by an amount in the ocean. This gives a total of 247ppmv to be transported away and brought back to the surface in each major cycle to continue to satisfy Henry's Law. A possible mechanism for transport is as follows: Cold water is heavier than warm water and sinks making lower depths colder. The pressure also increases with depth. Water, being a polar compound, bonds its molecules together to some extent, which helps to produce surface tension. $CO_2$ molecules, on the other hand, are not polar and move somewhat easier through the matrix. Molecules that move down however, have movement inhibited by lower temperature and higher pressure and get trapped so there is a net flow down. This prevents buildup at the surface during the absorption part of the cycle, and allows absorption to continue. What happens during release is a little harder to explain. A mechanism is described in this and other companion reports whereby buildup of atmospheric concentration is the result of wintertime thermal diffusion across the air/water interface of northern oceans. For release to continue a mechanism has to be found to replenish $CO_2$ in the water surface. It is hereby postulated that this is achieved by mixing of the upper layers of the ocean. Mixing overcomes the downward migration mechanism, acting effectively as a pump to provide the $CO_2$ required and the buildup mechanism draws it into the atmosphere.

## 5.3 Biomass intervention

Exponential release of $CO_2$ at the beginning of an interglacial period is stopped by intervention of the biomass. Climate control therefore has indeed depended on life, a concept discarded at the beginning of this report series. It is accepted in popular literature (e.g. Ericson, 1989), that it is life that has controlled the climate throughout the long history of the Earth. In Bell (2002) this idea was rejected in favor of cloud control in order to accommodate the idea that recent climate cycling began as a result of an oscillating control system. It now turns out that, although life is not normally required, without it the oscillating system would long since have spiraled completely out of control.

# 6 CONCLUSIONS

1. For the large thermal cycles of recent interglacial periods, the ocean temperature lags that of the rest of the environment by about 12,000yr, in agreement with previous work (Bell, 2002 and 2003).

2. The disparity of ocean and atmosphere temperatures, the result of the lag, determines the amount of cloud cover and that in turn is effective in causing exchange of $CO_2$ between atmosphere and ocean, one being the result of the other.

3. At the start of an interglacial period the release of $CO_2$ from the ocean is approximately a function of the fourth power of time, indicating the existence of aggressive feedback conditions. After a certain period, release changes from an exponential to a linear function of time, probably because of the depletion of northern ocean surface layers.

4. Linear release of $CO_2$ continues at about 10ppmv/1000yr, the rate probably being determined by the rate of transfer from the ocean depths. (For this to be true large amounts of $CO_2$ have to be transported to and from the ocean depths in each interglacial/glacial cycle to satisfy Henry's Law at the water surface. This in turn requires that migration of $CO_2$ in the ocean be, at certain times, the result of thermal and pressure diffusion and at other times, the result of concentration diffusion and ocean mixing).

5. $CO_2$ buildup is eventually stopped by biomass intervention. Continuing growth of northern forests then maintains the $CO_2$ concentration more or less constant and decreases atmosphere temperature allowing the ocean temperature eventually to catch up and surpass that of the atmosphere. After the temperature crossover there is negative disparity and $CO_2$ is reabsorbed into the ocean.

6. The biomass displaced by the glacial cycles contains the equivalent of about 67ppmv of $CO_2$. The carbon

dioxide contained by the total biomass may be judged, therefore to be the equivalent of about 670ppmv.

7. It is further confirmed that the ocean is the major source and sink for $CO_2$ and that climate change is the result of diffusion across the air/water interface, a process that occurs as a result of concentration and temperature difference.

Acknowledgments

Acknowledgment is hereby given to *Tellus* for kind permission to republish figure 1, the data also being adapted for figures 2 to 6. Thanks are expressed for the advice and encouragement offered by John C. Anderson.

References:

Barnola JM, Pimienta P, Raynaud D, and Korotkevich YS (1991) $CO_2$-climate relationship as deduced from the Vostok ice core: a re-examination based on new measurements and on a re-evaluation of the air dating. Tellus. 43B: 83-90.

Bell LG (2002) World ocean temperature lag time: an analysis based on glaciation data for the last two million years. Theor Appl Climatol 73 (3-4): 243-247.

Bell LG (2003) Ice Age mystery: a proposed theory for the cause of long term climate change. Theor Appl Climatol 74 (3-4): 235-244.

Bell LG (2005) Interpretation of $CO_2$ data with ocean as source and sink. In preparation.

Berner EK and Berner RA (2001) Carbon dioxide concentration and climate over geological times, Encyclopedia of Global Environmental Change, Ed Munn RE. Chichester: John Wiley and Sons, 1: 250

Boden TA, Kaiser DP, Sepanski RJ, Voss FW (eds.) (1994) Trends '93A, Compendium of Data on Global Change. ORNL/CDAIC-65, Carbon Dioxide Information Analysis Center, Oak Ridge National Laboratory, Oak Ridge, Tennessee, USA.

Ericson J (1989) Living Earth. Blue Ridge Summit PA: Tab Books: 16

Jacob DJ (1999) *Introduction to Atmospheric Chemistry.* Princeton University Press: Princeton.

Jouzel J, Lorius C, Petit JR, Grenthon C, Barkov NI, Katliokov VM, Petrov VM (1987) Vostok ice core: a continuous isotopic temperature record over the last climatic cycle (160,000 years). Nature 329: 403-408.

Porter SC (1989) Some geological implications of average quaternary glacial conditions. Quaternary Research. 32:245-261, Figure 1: b and c.

## Thermal Diffusion
### July 16, 2005

According to Henry's Law of Solubility the concentration of a solute in liquid is proportional to the partial pressure of the solute in the gas in contact. If solute is added to the gas in a closed container the concentration in the liquid obviously has to increase by diffusion across the surface tension barrier to create a new balance.

No consideration has been given to the possibility of an effect of the imposition of a temperature difference. It is hereby maintained that increasing the temperature of either medium, while the other stays the same has the effect of increasing the partial pressure or concentration of that medium. Increasing temperature would increase the momentum of individual solute molecules thus allowing them more easily to cross the barrier thus striking a new and different balance.

Henry's Law applies to an ideal gas, one that does not react chemically with the solvent. $CO_2$ does dissolve to some extent in water, partially breaking down into acid and base components. It is therefore perhaps not the best for demonstrating the temperature-difference effect, however it is easy to hand. It is also an important solute to study in view of current buildup in the atmosphere and how that might be affected by the interaction between atmosphere and ocean.

If you take a cold can of carbonated drink from the refrigerator and crack it open, the fizz is gradually but inevitably lost. The process is slow because air in contact being relatively warm and containing about 375ppmv of $CO_2$ provides some back flow. Now should one take another can from the fridge and pour it over fresh ice cubes from the freezer, it is observed that there is violent foaming and the fizz is lost in seconds. This is because the gas layer on the cubes is now relatively cold and $CO_2$ contained therein cannot compete with somewhat warmer molecules in the liquid. In this latter case temperature difference causes solute molecules to climb from a relatively dilute solution in the liquid to 100% $CO_2$ gas. Since temperature differences are small this demonstrates a potent thermal diffusion effect.

Analysis of data from NOAA observatory at Mauna Loa shows that the rate of $CO_2$ increase reaches a maximum in January then decreases to a minimum in August. It is argued by mainstream climatology that the January peak is the result of the decay of land bound biomass, delayed in its migration to this remote site far from land. There is however much evidence to refute this. Data obtained by container ships shows for example that the rate in January increases with increasing latitude whereas decay would surely go the other way. Could it be that cold winter air from Siberia causes release by diffusion from the ocean?

Wikipedia is an Internet encyclopedia making claims to contain information from encyclopedia from all countries and in all languages and yet there is no entry for thermal diffusion. Thermal diffusivity is mentioned but that has to do with the dispersion of heat whereas thermal diffusion has to do with the transport of matter caused by temperature gradients. TD is a well-known phenomenon in atomic energy because of the induced migration of hydrogen through zirconium alloy pressure tubes. Smaller atoms of hydrogen from corrosion by hot coolant migrate through the zirconium lattice, down the temperature gradient and precipitate as embrittling hydride in the tube wall.

Sawatzky A., CE Ells; Understanding Hydrogen in Zirconium, American Society for Testing Materials, 100Barr Harbor Drive, West Conshohocken, PA 19428-2959, AECL No. 12054

Thermal diffusion can occur in solids with atoms of different but similar size and in this circumstance smaller atoms can flow either way, up or down the temperature gradient, depending on the heat of transport. Interstitial vacancies have to migrate the other way. The phenomenon is not recognized, however, in gases or liquids, as strange as that may seem and as far as I know.

Borg RJ, GJ Dienes; An Introduction to Solid State Diffusion, Academic Press, 1988

It is observed (Barnola et al) that during the latter part of an interglacial period and the ensuing glacial stage $CO_2$ in the atmosphere decreases by about 95pmv. This is a lot of $CO_2$. It could not be absorbed by the biomass because the northern forests are retreating during this time. The product of decay of the northern biomass also has to be disposed of. The only possibility is that the ocean absorbs excess $CO_2$ from all sources. According to Henry's law the concentration of water in contact has also to decrease in proportion. This means that possibly as much as 250ppmv has to be transported from the surface layers to the ocean depths. For this to happen it seems likely that there is a thermal/pressure diffusion effect. The lower temperature and higher pressure trap $CO_2$ molecules that happen to move down thus creating a net flow. The absence of an accepted theory to explain this provides another reason to believe there is an uninvestigated area of science.

Barnola J. M.; D. Raynaud, C. Lorius and Y. S. Korotkevich, 1994. Historical Record from the Vostok Ice Core. *Trends '93A*: 7-10.

I am going through this exercise because this is an important area of science that is being completely overlooked. Perhaps you, a colleague or an acquaintance might be prepared to conduct some closely controlled experiments.

# Chapter 5
## Earth History

There happened to be a story in the news about the discovery of a period in the Earth's history when the ocean became frozen.  The "Snowball Earth" condition was believed to have lasted many millions of years before 550myr BP with even the ocean frozen over. It was only speculated that some portion of the ocean did not freeze because life managed to survive.  A few years later there was another report about the discovery of extensive volcanic activity in what is now Siberia that caused a drastic change in life forms. Events such as these are not satisfactorily explained in mainstream climatology and it was decided to attempt an explanation in accordance with the new theory being developed. Traumatic events, such as the above, as well as the disaster that ended the age of the dinosaurs, are said to have caused extremely high levels of carbon dioxide, 2000ppmv or more. Since calculations show that each ppmv represents about $8.2 \times 10^9$ tonnes, the concentrations speculated represent a tremendous amount, more than the $CO_2$ component of the biomass, and much more than could be the result of a local impact or volcanic activity. Such large amounts have only one source and that again is the ocean.

The pasted report was submitted to two journals but was not accepted for publication. In one case the editor was about to retire and his designated reviewer was reluctant to accept another report, so he had to give it a pass. In the other the excuse given was that there wasn't space for it in the current issue and "it falls below the novelty and importance necessary for further consideration".

# 5. Earth History with the Ocean as the Source and Sink for CO$_2$: Extenuation of a Theory on the Cause of Climate Change

## L. G. Bell

*Abstract*

   *The life-induced mechanisms of conventional climatology theory are found to have insufficient capacity to explain dramatic climate changes of the past. Information available is related to a new theory, which includes the idea that the ocean is the main source and sink for CO$_2$. The climate is usually maintained in a controlled state by the amount of cloud cover, the CO$_2$ concentration being automatically adjusted to offset extraordinary changes in the Earth's albedo. However if a continent moves into Polar Regions, control can be destabilized. On one occasion in the far distant past continental drift is thought to have caused a severe ice age, almost extinguishing life on Earth. Exponential release of CO$_2$ from the ocean was eventually triggered causing an extended period of hothouse conditions, which further jeopardized the survival of life. Normal control was eventually reinstated and CO$_2$ levels declined as continents drifted away from the poles and as the Sun's heat and the land area increased. At the beginning of the Pleistocene Epoch, the system became vulnerable because of continental drift, at some point inducing rapid climate change; the ocean temperature was affected and the control system started oscillating. Then, because of increasing variation in glaciation and CO$_2$ concentration, climate cycles generally increased in magnitude and duration. In this latter instance life has prevented the extreme cycles of the far distant past by limiting CO$_2$ variation.*

## 1. INTRODUCTION

   It is a commonly held belief that it is the actions of life forms that have controlled, or at least determined, the CO$_2$ concentration in the atmosphere over the long history of the Earth. It is claimed that increasing levels of CO$_2$ are caused by excessive volcanic activity and that decreasing levels are caused either by the deposit of coal, by increasing biomass (Berner et al, 2001), or by the action of plants (Berner, 1997).

   One objective of the present report is to examine the validity of the conventional life-control theory with biomass and other life-related mechanisms as source and sink for $CO_2$. A second objective is to apply the essence of the proposed new theory with the ocean as source and sink to what is known about the longer history of the Earth.

## 2. PART 1, $CO_2$ REPOSITORIES

### 2.1 Introduction

There are long-term carbon cycles involving various forms of life. Limestone, for example, is laid down over millions of years by the action of organisms in the ocean. Carbon, as $CO_2$, is then eventually released to the atmosphere by volcanoes after many more millions of years when the limestone deposits are subducted by continental drift, or by weathering when sedimentary rock is exposed to the atmosphere. Carbon may also be taken out of the environment by plants when fossilized remains are accidentally stored and become compacted to form coal.

The objective of this report-section is to test these factors as possible means of climate control using available factual evidence and the ocean-as-source-and-sink theory that is being developed.

### 2.2 Information

It is reported (Encyclopaedia Britannica, 1997) that the known reserves of coal are about $10 \times 10^{12}$ tonnes (only about $0.6 \times 10^{12}$ tonnes are considered to be recoverable). Since six kilograms of $CO_2$ are required to produce a kilogram of coal, this represents $60 \times 10^{12}$ tonnes of $CO_2$ taken out of the environment and deposited as coal. By rough calculation 1ppmv of $CO_2$ in the Earth's atmosphere is the equivalent of $8.2 \times 10^9$ tonnes of $CO_2$. Total coal deposit therefore represents a reduction of $60/8.2 \times 10^3 = 7,300$ppmv of atmospheric $CO_2$. (As an interesting aside, burning all of the coal considered recoverable would produce $7300 \times 0.1/60 = 439$ppmv. Dividing by 2, because only about half stays in the atmosphere (Bell, 2005b), gives an estimate of 220ppmv potential buildup)

It is also reported that most of the coal was deposited during a 100myr interval between 350myr BP and 250myr BP, starting about 10myr after the start of the Carboniferous Period and continuing on into the early part of the Permian. Smaller but significant deposits were made between 135 and 2.5myr BP.

Estimates have been made (Berner, 1997) of atmospheric $CO_2$ concentration using a "long-term carbon cycle model" based on life-induced storage mechanisms as source and sink. The graph of average estimates over time (Berner et al, 2001) is shown in figure 1, with $[CO_2]$ scale changed from $RCO_2$ to ppmv. It will be noted that $CO_2$ concentration is estimated to have decreased from a high of about 5600ppmv at 520myr BP to a low of about 300ppmv at 300myr BP.

It is found in a companion report (Bell, 2005a), that during 25,000yr of the interglacial period previous to the present one, about 162ppmv of $CO_2$ were released by the ocean, 95ppmv being retained in the atmosphere and 67ppmv being absorbed by the biomass. A judgement based on the latter figure is that the total biomass of the planet contains about 670ppmv of $CO_2$. (A means is found in a subsequent report to increase the 670ppmv estimate to 1220ppmv). These estimates may be on the low side because they do not agree very well with the generally accepted view (Jacob, 1999) of 4000PgC for the total biomass, which would be equivalent to about 1800ppmv should it be introduced to the atmosphere.

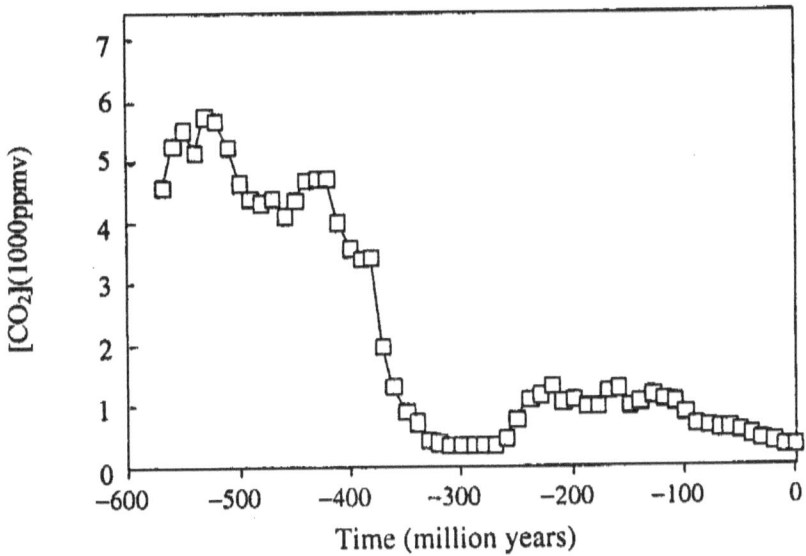

*Figure 1. Average atmospheric $CO_2$ over time as estimated (Berner, 1997) and as published by Berner et al (2001). The $CO_2$-concentration scale has been changed from $RCO_2$ to ppmv. Estimates were based on the idea that life processes control the concentration profile.*

## 2.3 Analysis and Discussion

Without the ocean as sink, it is assumed that proponents of the life-storage theory have to believe that the $CO_2$ level is reduced as in figure 1 from 5600ppmv at 520myr BP to 300ppmv or so at 300myr BP by the growth of the biomass. As noted above, the total biomass of the Earth is expected to be only that amount associated with 670 to 1800ppmv. It is not likely therefore that it could absorb the required 5300ppmv during this time or any large amount at any other particular time. This rules out the biomass as the effective major sink. Another strange result that comes from regarding biomass as the sink is that the minimum at 300myr BP falls right in the middle of the Carboniferous period when the atmosphere is supposed to have been exceptionally rich in carbon.

As calculated above, 7100ppmv of $CO_2$ in the form of coal was deposited over all time, most of it during the 100myr period around the end of the Carboniferous Era. If we assume that it was all deposited during this 100myr we have a deposit rate of only 71ppmv/myr. At one point in history, at least 152ppmv were exchanged between the ocean and the rest of the environment in only 25,000yr (Bell, 2005a). This may be taken as a minimum amount for the $CO_2$ that is circulated at all times, which would suggest an exchange rate of at least 6100ppmv/myr. It is unlikely therefore that 71ppmv/myr deposited as coal would have a noticeable effect on the $CO_2$ level. This essentially rules out coal as a major sink.

The argument for volcanoes not being the source for increasing $CO_2$ would be much the same as that for decreasing by coal deposit, if there were data on the amount and history of limestone deposits. Without that information we can still argue that the process of deposition, subduction, and release by volcanic activity is something that goes on continuously and would probably not have a noticeably different effect for any particular geological period. In unpublished work it is noted that there is no indication in ice core data profiles that volcanoes have any effect (Bell, 2005a). Considering the rapid exchange between atmosphere and ocean mentioned above, limestone deposits could therefore reasonably be ruled out as a source for increasing $[CO_2]$. It is hereby theorized that it is what is happening at the surface of the ocean that governs and outweighs the effects of incidental storage mechanisms.

## 3. PART 2 - HISTORY REVISITED

### 3.1 Introduction

The objective of this report-section is to reexamine Earth history with the ocean taken as being the significant source and sink for $CO_2$.

### 3.2 Information

### 3.2.1 Historical events

Very severe ice-age conditions are known to have existed for many millions of years prior to 550myr BP. The continent of Rodinia, making up most of the land surface area at the time, had drifted into position at the south magnetic pole. Evidence of the Veranger glaciation, as it is called, has led investigators to believe that temperatures continued to fall until the ocean was almost completely frozen (Hoffman et al, 1993). Life on Earth should have been extinguished by the "Snowball Earth" conditions, but since it was not some area of the ocean is assumed to have remained ice-free. It has been determined (Crowley et al, 2001) that at least 4500ppmv of $CO_2$ in the atmosphere would have been required to overcome the albedo feedback cooling and thus bring the ice age to an end.

Following the ice age there was an extended period of extremely warm climate lasting about 200myr. Although extremely warm, certain areas had to remain cool enough for life to survive and even flourish (Runnegar, 2000). Toward the end of this time there is evidence of minor glaciation on the polar continent (Appenzeller, 1993) and of the reintroduction of land plants (Berner, 1997). Evidence is then found of another severe ice age that occurred with the continent of Pangea, somewhat smaller than Rodinia, in position at the south magnetic pole (Hyde et al 1999). The period of glaciation in this instance supposedly began about 200myr after the end of the Veranger glaciation and lasted about 100myr centered on the Carboniferous Period at about 320myr BP.

A cataclysmic event occurred at the Permian-Triassic (P-T) boundary, 251.4myr BP, around the end of the Carboniferous Age (Renne et al, 1995). A heavy mass of lava was exuded from the Earth at what is now Siberia and very large amounts of $CO_2$ were released into the atmosphere.

Another cataclysmic event occurred as a result of an asteroid impact at the Cretaceous-Tertiary boundary (K-T) to end the age of reptiles about 65myr BP (Beerling et al, 2002). Once again large amounts of $CO_2$ were injected into the atmosphere, the level estimated to be still as high as 2300ppmv, 10,000yr after the event.

### 3.2.2 Recent Data

At the start of an interglacial period, atmosphere temperature leads that of the ocean thus producing clear skies and the amount of $CO_2$ accumulated in the atmosphere increases with time (Bell, 2005a). It is believed this happens because diffusion across the air/water interface is from warm to cool. Means is thus provided for release of $CO_2$ from northern oceans by clear skies, which reduce winter air temperature in the far north. (Recent data shows that this is currently happening in the northern Pacific (Bell, 2005b)). The increase is eventually stopped by biomass intervention. Soon after the $CO_2$ increase is halted, the atmosphere begins to cool and the ocean temperature, which lags that of the atmosphere by about 12,000yr (Bell, 2002), is allowed to "catch up" to that of the atmosphere. After this second temperature crossover, with the ocean now warmer than the atmosphere, the climate becomes cloudy, the biomass starts to decay and $CO_2$ is reabsorbed into the ocean. The atmosphere temperature declines while the ocean temperature continues to increase, reaching its maximum 30,000yr or so into the interglacial period. Conditions thenceforth remain marginally cloudy for the remainder of the interglacial period and the ensuing glacial stage because the ocean, lagging atmosphere, stays relatively warm during this time. A glacial stage will only end when the ocean temperature, by whatever means, is induced to drop below that of the rest of the environment to once again provide the relatively clear skies required to start the ice melting and to release $CO_2$ (Bell, 2003).

### 3.3 Analysis and Discussion

### 3.3.1 Veranger glaciation

The Rodinia ice age probably began, the result of heat lost by glaciation, as the large continent gradually moved into position at the south magnetic pole. Because of ocean temperature lag, as noted above, marginally cloudy conditions should prevail during a glacial period causing continually reducing $[CO_2]$ and continually decreasing atmosphere temperatures. The period of glaciation should end when the ocean starts to freeze, causing the ocean temperature to drop below that of the atmosphere. For the Veranger glaciation the ocean temperature could not fall low enough because too much heat was being lost by reflection from the very large glaciated continent. This is why the glacial period could last so long.

The ocean temperature could not fall low enough to produce clear skies, however, as the ocean became frozen, evaporation would have been reduced because of the decreasing amount of water surface. The resulting clearer skies provided a means of low temperature control. Some $CO_2$ was released from the patch during the night and in winter because of cold air in contact. If too much was released the water patch would get too big and clouds would increase thus causing absorption. A certain level of $CO_2$ in the atmosphere was thus maintained with just the right size of exposed water to release and absorb the required $CO_2$ to maintain the stable state and thus produce what might be called a low temperature ice age. Life managed to survive because of the continuing existence of open water.

Something happened to end the Veranger glacial regime. A piece of the continent could have broken away and started drifting northward, thus producing a continental arrangement much like what exists today. This would provide cooler air over the southern part of the patch at certain times of the year and trigger uncontrolled exponential release. Since life on land was certainly almost extinguished by the severe ice age, the biomass could not intervene.

It was found in previous work, (Bell, 2005a) that at the beginning of the previous interglacial period, exponential release changed, after a relatively short time, to a linear rate of about 10ppmv/1000yr. This means that about 450,000yr would have been required to increase the $CO_2$ concentration to the 4500ppmv required by Crowley et al (2001) to melt the ice. The formula applies to the current continental configuration and a situation whereby the whole ocean is exposed to partial sunlight. With only a small patch of water exposed and a different configuration, the rate of release might have been different. However it was deemed in Bell, (2005a) that, in this circumstance, the linear rate of increase is determined by the rate of transfer of $CO_2$ from the ocean depths, which would not likely be affected by the particular arrangement.

3.3.2 Hothouse period

It is unlikely that 4500ppmv would have been required to melt the ice. Heating of the planet is dependent on time as well as $CO_2$concentration, and since it would take such a long time for the level to rise, the amount required would not be as great as that estimated with some other means of $CO_2$ increase. Whatever amount required would not have remained for long in the atmosphere, because once the ocean was ice-free, greenhouse warming would heat the atmosphere and the excessively high level of $CO_2$ would diffuse back into the cold ocean. As the $CO_2$ concentration dropped, disparity between ocean and atmosphere temperature would also drop giving time for the ocean temperature to catch up and perhaps stabilize the $CO_2$ level somewhere around 1000ppmv. That is not to say that the ocean and atmosphere would have reached the same temperature. It might be called the required effective relative temperature. Balance would occur when the ocean reached a temperature at which evaporation was equal to condensation and release of $CO_2$ was balanced by absorption. In recent climate cycling it is found (Bell, 2003) that increasing $CO_2$concentration causes clearer skies, however, with the ocean temperature following behind, there must be a point at which excessive evaporation causes cloud formation, perhaps in the stratosphere. This may have been what intervened in place of increasing biomass.

There was a danger of slipping back into major glaciation mode, however this did not happen. At 1000ppmv it is assumed that the average temperature for the ice-free zone would be in the range 25 to 30°C. This would have created a very humid environment. As described below, a control mechanism has to adjust the $CO_2$ concentration to offset heat lost by a polar continent. Perhaps control was achieved by means of stratospheric cloud cover. Clouds would effect control by reflecting heat away, by moderating winter temperatures (and therefore limiting the release of $CO_2$) and reducing the effectiveness of greenhouse warming. It is possible that a high temperature controlled state may have reigned for a considerable period of time. For easy reference this time will be called the Hothouse period.

After about 200myr with conditions as described something happened to disrupt the status quo. Perhaps another piece of the southern continent moved away from the southern continental mass causing the $CO_2$ concentration and atmosphere temperature to spiral down. In addition it is possible that the hothouse climate began to moderate because of the emergence of rooted vascular plants (Berner, 1997), not because of the $CO_2$ they consumed, but rather because of the change in albedo offered as the plants spread over bare rocks and sand. This is discussed further below.

### 3.3.3 Pangean ice age

By whatever means the hothouse period ended, however there was a smaller, but significant, continent now called Pangea, still in position at the South Pole and the world reverted to what might be described as warm ice-age conditions with $CO_2$ perhaps around 450ppmv. Climate control during this Carboniferous age was managed by the amount of low-level rather than stratospheric cloud as in earlier times.

After 20myr or so of this Pangean ice age, major forests began to proliferate in the carbon rich environment, presumably on land farther north, leaving records in the form of coal deposits. During this time the $CO_2$ concentration remained fairly high by our standards. The reason for this is that there was still a large glaciated continent at the South Pole reflecting heat away, thus to maintain a stable state as explained below, a high level of greenhouse gas was required to retain heat.

The cycling of the recent Pleistocene ice age began as a result of an oscillation of the cloud climate-control system as Antarctica gradually moved into position at the South Pole (Bell, 2002) and, more importantly perhaps, as Greenland moved to its present position about 2.4myr BP (Souches, 1997). Once started, period and amplitude of climate cycles were increased by many factors, one being an induced variation in atmospheric $CO_2$, but probably the most important being the presence of two large continental land masses, not at the opposite pole but close to it, providing accommodation for periodic glaciation. In the case of the Carboniferous ice age there was no opportunity for a control system to start oscillating. Pangea was already in position when the environment gradually cooled. There was also no large landmass near the opposite pole. In addition the land area was relatively small meaning that the ocean was larger than it is now and isolated from the main body of land at the South Pole. This would also help to prevent cycling. It seems quite likely therefore, that the ocean cooled and glaciers spread over Pangea, the planet easing into a warm ice age without inducing major instability.

When the Carboniferous ice age ended, instability would also have been unlikely because the $CO_2$ level was again decreasing. There was probably a peaceful transition to an ice-free stable state as Pangea gradually moved away from the south magnetic pole.

### 3.3.4 Dinosaur age and beyond

All during the time of the dinosaurs, from the end of the Carboniferous Age to the catastrophe that caused their demise, and from then until the beginning of the Pleistocene Epoch there was no continental landmass, at or near, the poles. As a result of this there was no major glaciation to interfere with climate control.

There were however two catastrophic events, major volcanic activity at 251.4myrBP and an asteroid impact at 65myrBP, both deemed to have introduced large quantities of $CO_2$ into the atmosphere. How this would happen with the ocean as source is as follows: each of these events would have thrown large quantities of fine particulate matter into the stratosphere thereby causing a period of very cold climate. The cold air would draw $CO_2$ out of the relatively warm ocean causing a buildup in the atmosphere. Eventually the dust particles found their way back to the surface, greenhouse warming then warmed the atmosphere and the $CO_2$ was reabsorbed into the ocean. Though problematic for the survival of life forms these events only cause a momentary hiatus in the normal cloud-cover/$CO_2$ control.

It is noted that the $CO_2$ concentration increased in a relatively short time in both of these two catastrophic events, but then remained high for a long period of time. This can be explained by means of the proposed theory.

At both of these events the atmosphere was cooled by dust in the stratosphere and $CO_2$ diffused out, depleting surface layers of the ocean. So, to replenish, it diffused up from the depths, a relatively rapid process. When the dust settled out the atmosphere heated but now to get the ocean to accept such a large amount, it had to diffuse back down to the ocean depths with the help of pressure and temperature gradients but against a concentration gradient; something that could take much longer.

### 3.3.5 Profile over history

The supply of heat is mainly from the Sun. To provide conditions for life it is necessary to reduce and regulate this input. According to the proposed theory cloud cover achieves reduction by its effect on the Earth's albedo. The amount of cloud cover is mainly determined by the temperature of the ocean and, since that cannot be changed rapidly, cloud cover has a certain capacity for regulating the climate as well as reducing heat input. This has been effective for long term control over most of Earth history. Short-term control, on the other hand, was achieved by variation of $CO_2$ concentration. If the temperature of the atmosphere should increase for some reason diffusion into the relatively cool ocean would decrease $CO_2$ in the atmosphere and normal temperature be thus restored.

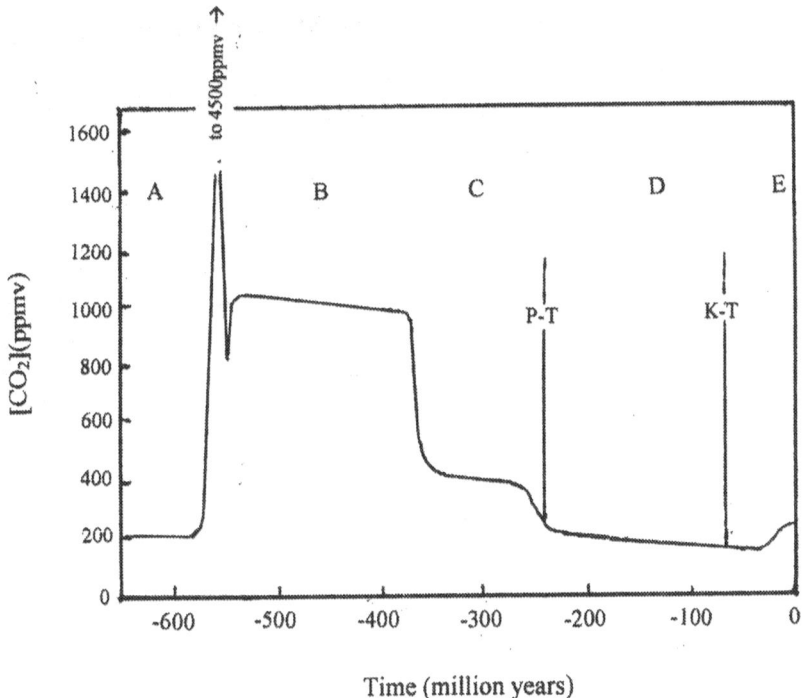

Figure 2. *Atmospheric $CO_2$ over the last 600myr as the climate is influenced by major events, based on the idea that the ocean acts as source*

*and sink for $CO_2$. Distinct periods are indicated by letters as follows: A: Rodinian (Veranger) cold ice age, B: Hothouse inter-ice-age, C: Pangean (Carboniferous) warm ice age, D: Controlled climate with no polar continent, E: Antarctic ice age of the Pleistocene Epoch. Catastrophic events are indicated at the P-T boundary, 241myrBP, and at the K-T boundary, 65myr BP.*

Heat input has to balance heat loss in order to maintain steady state. By means just described the control system automatically adjusts the $CO_2$ level, lower heat input (eg. increasing albedo) requiring more $CO_2$, and higher heat input less $CO_2$. Input was very low during the Hothouse period, heat being reflected away by cloud cover and the large bare southern continent, glaciated or not. Input increased somewhat as plants spread over the land in the Hothouse period but was still low during the Carboniferous ice age, then increased again in step function at its ending. During the stage just prior to the Pleistocene, including the age of the dinosaurs, and for all time actually, there was the continuing extra bonus of heat input as the Sun's power increased and as the land surface area increased while that of the ocean declined. All during the time of cloud-cover control the $CO_2$ decreased, reacting to increasing heat input. After the start of the Pleistocene Epoch the average $CO_2$ level is expected to have increased again to some extent, because Greenland and Antarctica are where they happen to be and because of extra heat lost during intermittent periods of glaciation. Figure 2 shows proposed, and mainly guessed at, levels of $CO_2$ throughout the last 600myr with ocean as source and sink.

### 3.3.6 Glaciation effects

As described above, heat lost by glaciation has to be compensated by heat input. This is achieved automatically by cloud cover with its ability to change the albedo as well as the atmospheric $CO_2$ concentration. When continents gradually move toward or away from Polar Regions the control system has a certain ability to offset the change in heat loss from glaciation by changing $CO_2$ concentration. Indications are that this is what happened over most of the Earth's history since 550myr BP. There is one occasion when it did not.

When the continental land areas of North America and Eurasia moved to their present positions the stage was set for an unusual scenario because large areas were provided to accommodate glaciation, not at the North Pole but close to it. Sometime during the migration of Antarctica to the South Pole and of Greenland to its present location, control became unstable. The temperature of the ocean began to oscillate opposite to the atmosphere causing alternating periods of clear and cloudy conditions. Eventually the cycling became more severe with cycles being enhanced by glaciation and excessive variation of atmospheric $CO_2$. Essentially the problem is that the long-established control system has been compromised by the continental configuration. In the distant past when $CO_2$ was increased, skies got clearer, the atmosphere temperature increased and the $CO_2$ was reabsorbed. Now adding $CO_2$ causes general heating but northern winters become colder causing more $CO_2$ to be released. It could be said that this instability is essentially what has caused the cycling of the Pleistocene Epoch. Life is now an important factor, because without it the system would be vulnerable to runaway in both directions. It is only because of biomass intervention that conditions have been kept within moderate bounds.

## 4. PROPOSED CONCLUSIONS

4.1 Life induced storage and release, such as coal deposit, volcanism, biomass growth and decay, etc. have insufficient capacity to have acted as effective sources and sinks for atmospheric $CO_2$ at any time in history.

4.2 The ocean as source and sink for $CO_2$, with interface diffusion as the operative mechanism, fits reasonably well with what is known about events over the course of history back as far as 600myr BP.

4.3 Ocean temperature has a moderating effect because it determines the amount of cloud cover and that regulates heat from the Sun. It also to some extent determines the amount of $CO_2$ in the atmosphere because the temperature of the atmosphere controls diffusion across the interface between atmosphere and ocean.

4.4 $CO_2$ concentration seems to maintain control through gradual change such as that induced by continental drift and provides a means of recovery soon after catastrophic events.

4.5 It is theorized that climate control requires heat input to have an inverse effect on the $CO_2$ level. The latter has therefore decreased over most of Earth's history since 550myr BP, as glaciation related albedo cooling decreased, the result of continental drift, as the land area increased, and as the Sun's power increased.

4.6 A unique situation was set in motion as a result of the present continental configuration, large land areas having moved near the Arctic Circle. Subsequent movement of Antarctica and Greenland is believed to have caused the normally effective control to begin oscillating with temperature cycles being supported and enhanced by glaciation and atmospheric $CO_2$ cycles.

4.7 Indications are that the long-term control by $CO_2$ has been compromised in the Pleistocene by unusual winter climate in the far north.

4.8 Climate cycling of the Pleistocene Epoch is moderated by intervention of the biomass

Acknowledgements

Thanks are expressed for the advice and encouragement offered by John C. Anderson and Morley Thomas.

References

1.  Appenzeller, T., Searching for clues to Ancient Carbon Dioxide. *Science*, 259: 908-909, 1993

2.  Beerling , D.J., Lomax, B.H., Royer, D.L., Upchurch Jr., G.R., and Kump, L.R., Proceedings of the National Academy of Sciences, USA. Vol. 99, No 12: 7836-7840, 2002

3.  Bell, L.G., World Ocean Temperature Lag Time: An Analysis Based on Glaciation Data for the Last Two Million Years, *Theor Appl Climatol*, 73 (3-4): 243-247, 2002.

4. Bell, L.G., Ice Age Mystery: A Proposed Theory for the Cause of Long Term Climate Change. *Theor Appl Climatol,* 74 (3-4): 245-253, 2003.

5. Bell, L.G., Ocean/Atmosphere Temperature Disparity Effects: An Analysis of $CO_2$ and Temperature Data for the Previous Interglacial Period. 2005a (in preparation).

6. Bell, L.G., Interpretation of $CO_2$ Data with the Ocean as Source and Sink. An extenuation of a theory on the cause of climate change. 2005b (in preparation)

7. Berner, E.K., and Berner, R.A., Carbon Dioxide Concentration and Climate over Geological Times. Munn R.E., (Ed.), *Encyclopedia of Global Environmental Change,* John Wiley and Sons, Chichester, 1: 249-254, 2001.

8. Berner, R.A., The Rise of Plants and their Effects on Weathering and Atmospheric $CO_2$. *Science,* 276: 544-546, 1997.

9. Crowley, T.J., Hyde, W.T., and Peltier, W.R., $CO_2$ Levels Required for Deglaciation of a "Near Snowball" Earth, *Geo Phys Res Lett,* 28: 283-286, 2001.

10. *Encyclopaedia Britannica,* 3: 406 and 19: 595, 1997.

11. Hoffman, P.F., Kaufman, A. J., Halverson, G. P., and Schrag, D. P., A Neoproterozoic Snowball Earth, *Science,* 281: 1342-1346, 1998.

12. Hyde, W. T., T. J. Tarazov. L., and Peltier, W. R., The Pangean Ice Age: Studies with a Coupled Climate-Ice Sheet Model, *Climate Dynamics,* 15: 619-629, 1999.

13. Jacob, D.J., *Introduction to Atmospheric Chemistry.* Princeton University Press: Princeton, 1999.

14. Renne, P.R., Z. Zichao, M.A. Richards, M.T. Black, and A.R. Basu, Synchrony and Causal Relations Between Permian-Triassic Boundary Crises and Siberian Flood Volcanism, *Science* **269**: 1413-1413, 1995.

15. Runnegar, B. Loophole for Snowball Earth, Nature, 405: 403-404, 2000.

16. Souches, R. The buildup of the ice sheet in central Greenland, *Journal of Geophysical Research*, 102(C12): 26317-26323, 1997.

This analysis covering all of history provides further confidence in the new theory with its tenets of cloud cover control and release and re-absorption of $CO_2$ by thermal diffusion when conditions permit. A significant finding, as far as predicting the future is concerned, is that $CO_2$, at one point in history went from very low to very high values in a relatively short period of time. Moreover this happened at a time when there was a continent at one of the poles.

In this and other reports in the series it is mentioned that 1ppmv of $CO_2$ in the atmosphere is calculated to be 8.2 billion tonnes. Perhaps for the record I should explain how this was done.

- My world atlas shows Earth to have a surface area of $197 \times 10^6$ square miles, which works out to be $7.88 \times 10^{17}$ square inches.
- The total weight of the atmosphere is thus obtained by multiplying the area in square inches by the atmospheric pressure of 15psi. This gives $1.18 \times 10^{19}$ lb as total weight. Changing from pounds to kilograms gives a total atmosphere weight of $5.36 \times 10^{18}$ kilograms. Changing this to tonnes, and dividing by $10^6$, gives a value of $5.36 \times 10^9$. tonnes, or 5.36 billion tonnes for 1ppm.
- $CO_2$ is heavier than air and, to get its weight, it is necessary to change from ppm to ppmv. From Avogadro we know that the weight of

a certain volume of gas is proportional to the atomic number. We therefore multiply 5.36 by 44, the atomic number of $CO_2$, and divide by 28.8, the atomic number of air, 20%O and 80%N. This gives a weight of 1ppmv $CO_2$ to be 8.2 billion tonnes.

PgC is the measure of the amount of carbon and is in units of $10^{15}$ grams. Multiplying 8.2 for 1ppmv by 12 and dividing by 44 gives an amount for carbon alone as 2.24 PgC, (the powers of 10 cancel each other).

The amount of carbon in the biosphere is estimated by climatolgists, and mentioned a few times in this book, as 4000 PgC. This works out to 1785ppmv, or about 1800ppmv, as an equivalent amount of carbon should it be introduced to the atmosphere as $CO_2$.

# Chapter 6
## $CO_2$ Data Interpretation

It was assumed, up until the early 1950s, that $CO_2$ released into the atmosphere would dissolve in the ocean. It was thus a surprise to climatologists, when Charles Keeling devised means for measuring the $CO_2$ concentration, to find that it had increased above what had been generally accepted. Moreover a station established on Mauna Loa, Hawaii found that there was a continuing increase with time and an annual variation of several parts per million.

Since that time there has been a very large amount of data collected, on Japanese container ships and at Mauna Loa and other remote sites. The data is readily available on the NOAA web site and in journal reports but no one has undertaken an analysis. Those investigators suffering isolation and uncomfortable conditions have to be commended for their dedication. It must be a disappointment to them that no one makes use of the data.

The report pasted here is my effort to derive meaning. It was submitted to several journals and violently rejected by all of them. Perhaps a reader can tell me why. Is it really poor science or could it be fear of the implications to our future wellbeing?

# 6. Interpretation of $CO_2$ Data with the Ocean as Source and Sink

## L.G. Bell

*Abstract*

*It is maintained that the ocean, rather than the continental biomass, acts as the main source and sink for $CO_2$, and that diffusion across the air/water interface is the controlling process. The theory is tested against $CO_2$ and $\delta$ $^{13}C$ data from shipboard sampling and remote stationary observatories. Increasing annual fluctuation with increasing latitude and other aspects of the data are examined as interface diffusion is affected by proximity, cloud, ice, altitude, temperature, isotope selection and sunlight variability. In agreement with a diffusion theory, it is found that air temperature has the greatest influence, $CO_2$ being released from the ocean in high northern latitudes in winter and reabsorbed in summer. The finding, that carbon isotope ratio $\delta$ $^{13}C$ values follow in opposition to the $CO_2$ data, supports the diffusion theory and suggests that there is a preference for $^{12}C$ over $^{13}C$ in the diffusion across the air/water interface in both directions. In addition evidence is found for annual release and absorption of $CO_2$ by polar ice, a phenomenon also attributed to a thermal diffusion effect. The rate of $CO_2$ concentration change over several years, the result of global warming, increases with latitude, a significant finding since it suggests that an excess of $CO_2$ release from the ocean is supporting or adding to global buildup.*

## 1 INTRODUCTION

Generally accepted theories of climatology, as described for example in Jacob, (1999) maintain that the biomass is the main source and sink for $CO_2$ and that variation in measured concentration is related to this phenomenon. Minor transfer of $CO_2$ between the atmosphere and ocean can occur in conventional theory by means of a chemical process. Equations have been developed for a buffered acidification by dissolved $CO_2$ as it is influenced by the pH of surface seawater. Some transfer across the boundary from air to water is expected to occur but not from water to air because most of the $CO_2$ in seawater is in the form of $HCO_3^-$.

There is also in conventional theory a concept called radiative forcing, which is used to explain the increase in $CO_2$ levels in interglacial periods. Presumably change in insolation due to the Earth's precession and orbit eccentricity cycles forces a change in atmospheric $CO_2$ concentration, however no mechanism is described that would implement the change.

This is one of a series of reports that proposes a new approach to the cause of climate change. Tenets of the new theory are that a certain amount of molecular $CO_2$ remains in solution in seawater, that the ocean is the main source and sink and that transfer of $CO_2$ occurs by diffusion in either direction across the air/water interface. The direction and magnitude of the flux is determined by a difference in concentration and/or temperature of the two media.

In an earlier report (Bell, 2003), examination of ice-core $CO_2$ and temperature data for the previous interglacial period led to the conclusion that $CO_2$ is released from the ocean when skies are relatively clear and reabsorbed when relatively cloudy. Picking up the idea from conventional theory it was stated that sunlight heating of the surface of the ocean causes release, and that blocking access by cloud cover causes absorption. In solids, diffusing atoms may be induced to flow either up or down an imposed temperature gradient depending on the heat of transport (Borg et al,). In the case being considered, it is proposed that the surface tension of the water surface provides a partial barrier. Molecules of $CO_2$ in the warmer medium, whether air or water are more likely to have sufficient energy to penetrate the barrier. A net flow from warm to cool could therefore be expected. This is why it was assumed that sunlight heating of the water surface would cause release (and thus support, to some extent, the idea of radiative forcing). However, it is lately surmised that sunlight heats the air as much as or perhaps more than, the water, which would if anything cause the opposite effect. As a result, as will be shown, it now has to be accepted that it is *air temperature* rather than the amount of sunlight that is the determining factor. This changes release from summer to winter and absorption from winter to summer, but otherwise the theory appears to remain intact. Promotion of any new theory requires that it be tested against collected data and other known facts and, if possible, be shown to offer a preferred interpretation. This is the purpose of the present study.

Daily, and sometimes hourly, $CO_2$ concentration measurements are being made at various sites around the world to monitor the increase attributed to the burning of fossil fuel. Sites for monitoring stations were chosen, for the most part, far from land so that data would not be influenced by local sources. With ocean as source and sink mid-ocean sites are perhaps not the right choice for avoiding local influence. At these sites, if the theory holds, both source and sink are right at hand. Although the choice of sites may not have been the best for the purpose intended, the data obtained might prove to be useful to test the proposed theory.

## 2 INFORMATION

### 2.1 Variation with latitude

In 1953, C. D. Keeling developed a method of accurately measuring the amount of $CO_2$ in air samples and, at his instigation, measurements have been made since 1958 at Mauna Loa Observatory in Hawaii (Keeling, 2001). The famous Keeling curve (figure 1) shows that the average concentration and the rate of concentration change, have both been increasing ever since measurement began. In addition to the increase in average concentration, there is a noticeable annual variation of about 6ppmv with maximum occurring in May and minimum in September.

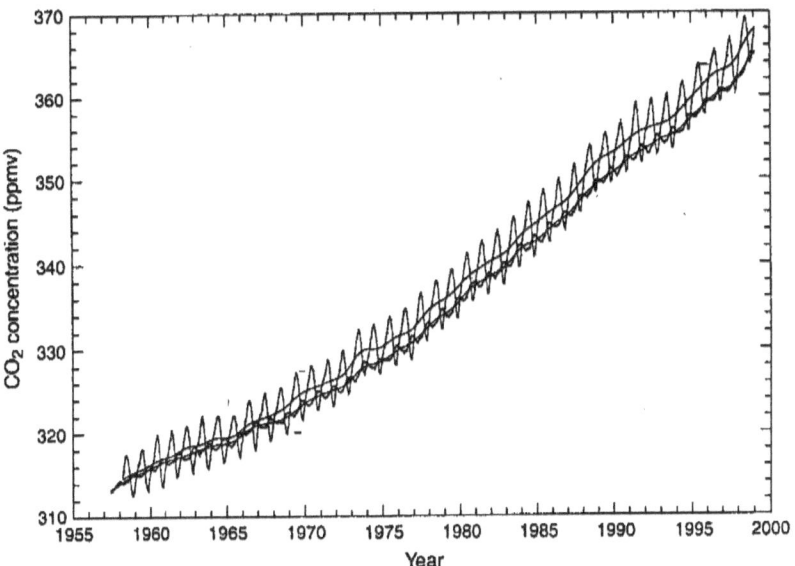

*Figure 1. Atmospheric $CO_2$ measurement data obtained at Mauna Loa showing an increase in concentration with time and an annual sawtooth variation of about 6ppmv. The corresponding data for the South Pole station are also shown. The South Pole shows an annual variation of*

*about 1ppmv and a current lag in average increase cf. Mauna Loa of about two years. The graph was copied from Higuchi et al (2001).*

There are four monitoring sites, now controlled by the Climate Monitoring & Diagnostics Laboratory (CMDL) in Boulder Colorado, more or less on a line from north to south at mid Pacific. Besides the one at Mauna Loa at about 18 °N, there is one at Point Barrow, on the Alaska north shore at about 70°N, one at American Samoa at 15°S, and one near the South Pole on Antarctica. All show a $CO_2$ increase over time. The rate of increase is found to vary with latitude as will be discussed below and the annual variation also changes from site to site. At Point Barrow the annual variation is up to 15ppmv, at Samoa there is no discernable variation and on Antarctica there is a barely noticeable cycle of about 1ppmv. The latter is out of phase with northern variation, in this case with maximum occurring in October and minimum in February. The Antarctic data are included in figure 1.

A very ambitious study has also been undertaken using air samples collected by Japanese container ships, one set following the great circle between Seattle and Tokyo, another between San Francisco and Tokyo and yet another tracing a course between Tokyo/Yokohama and Sydney/Melbourne, Australia. The results published by Morimoto et al, (2000) combined with CMDL data show that there is a uniform increase in annual variation in $CO_2$ concentration with increasing latitude. Since readings from container ships were taken at both ends of the great circle routes, they also show that there is very little change in concentration at any moment in time along any particular latitude across the width of the Pacific Ocean. As with Samoa, samples taken by ships travelling south along the eastern coast of Asia, showed no discernable annual variation in $CO_2$ concentration once they had crossed the equator. Some but not all, curves for both Morimoto and CMDL are shown in figure 2 for the years between 1984 and 1992. The results are reasonably consistent, but with stationary site measurements showing slightly lower $CO_2$ concentration than those for shipboard sampling (for this latter point see figure 4a).

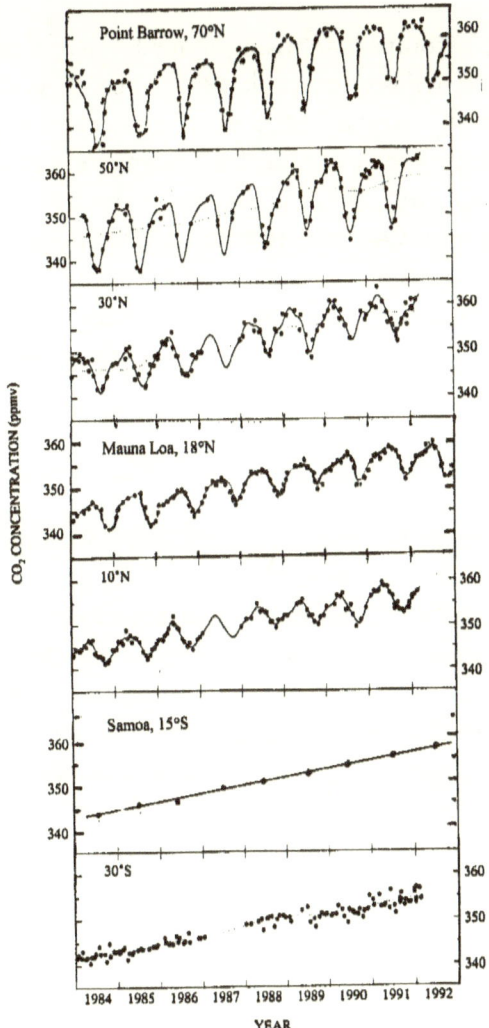

*Figure 2. CO$_2$ concentration data for the years 1984-1992 obtained using samples collected by Japanese container transport ships (Morimoto et al, 2000), interspersed with data for the same years obtained at CMDL stationary sites (Conway et al, 2003). Data points for Samoa are average for the year since monthly readings do not wander from the line. Change in annual variation is fairly consistent for all sampling locations, which provides evidence for variation being dependent only on latitude.*

2.2 Isotope selection

$^{13}$C is a stable isotope comprising about 1.1% of elemental carbon, the rest being $^{12}$C (Masterton et al, 1997). The $\delta^{13}$C ratio represents deviations from this standard. At present the ocean has a $\delta^{13}$C value of about 1 mil$^{-1}$ whereas the value for the atmosphere is about -8 mil$^{-1}$ and for the land biosphere, -25 mil$^{-1}$ (Higuchi et al, 2001).

In addition to $CO_2$ concentration, Morimoto et al (2000) also measured the fraction of $^{13}$C in their samples with the expectation of using the data to distinguish between the atmosphere/ocean flux and atmosphere/terrestrial-biosphere flux. The carbon isotopic ratio $\delta^{13}$C defined as:

$$\delta^{13}C = \lceil (^{13}C/^{12}C) \text{ sample } - 1 \rceil / \lfloor (^{13}C/^{12}C) \text{ standard } \rfloor 10^3,$$

was found always to be negative and to vary annually with a similar shape of curve, but opposite to that of the $CO_2$ concentration (Figure 3). A higher value of $\delta^{13}$C indicates a greater proportion of $^{13}$C. When the $CO_2$ is at maximum during the annual cycle the $\delta^{13}$C ratio is at minimum and vice versa. The $\delta^{13}$C maximum and minimum values of annual cycles also change in a regular fashion with increasing latitude, opposite to the change in $CO_2$ concentration. In addition the average $\delta^{13}$C value decreases with time whereas the $CO_2$ concentration increases. (Should one photocopy figure 3, fold the two graphs over each other and observe backlighted, it will be seen that the two curves have almost exactly the same shape and a similar trend).

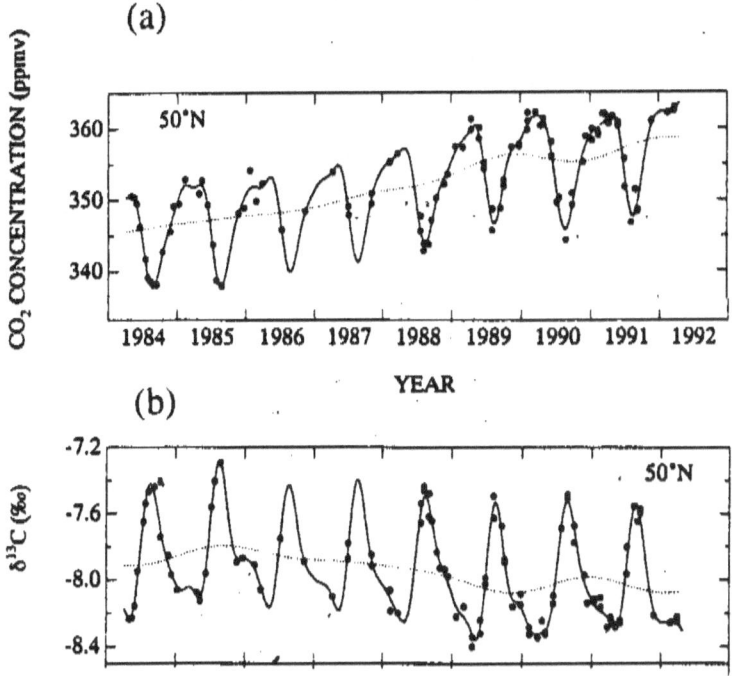

*Figure 3. CO$_2$ concentration and corresponding $\delta^{13}C$ values from Morimoto et al, (2000) for 50° north latitude. The $\delta^{13}C$ curve rises and falls directly opposite to that of CO$_2$ concentration. This provides further support for the interface diffusion theory.*

The maximum and minimum values of CO$_2$ concentration and for $\delta^{13}C$ value for one particular year, 1988, were plotted against latitude in figures 4a and 4b respectively. Note that the maximum CO$_2$ concentration, achieved in April, increases over the whole range of latitude, while the corresponding value for $\delta^{13}C$ declines. Above the equator, sine 0.0, minimum values for CO$_2$, achieved in August, decrease with latitude, while the corresponding $\delta^{13}C$ value increases.

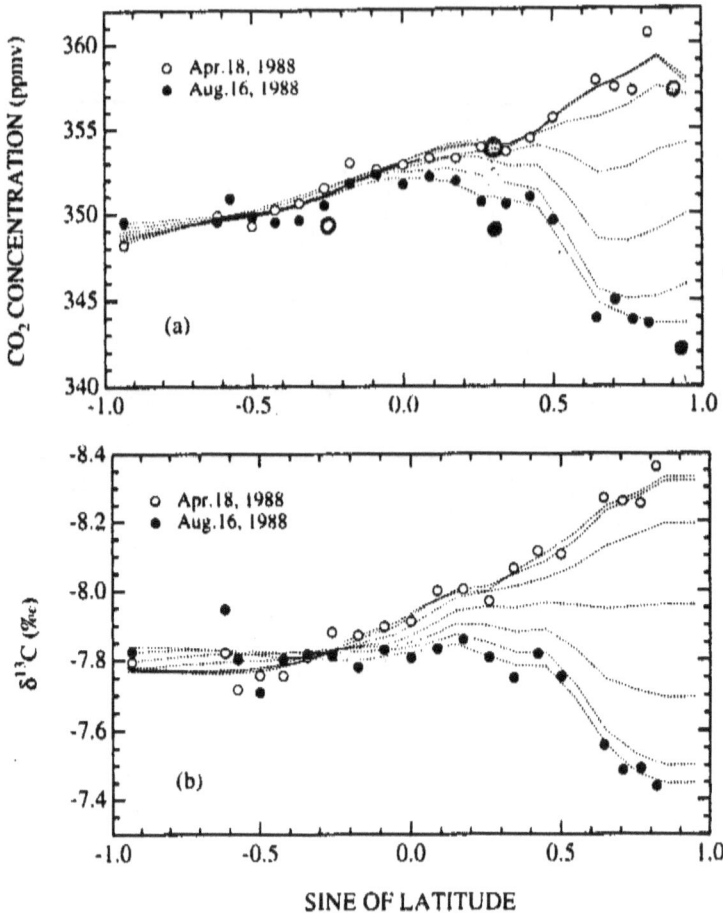

*Figure 4. Annual maxima and minima for both $CO_2$ concentration and $\delta^{13}C$ values as a function of the sine of the latitude angle. Both factors are observed to change consistently with latitude. Data comes from Morimoto et al, (2000) with some points for $CO_2$ from CMDL sites included using larger dots and circles.*

2.3 General aspects of $CO_2$ distribution

The average $CO_2$ concentration at the South Pole lags behind that of sites in the Northern Hemisphere. When compared with Mauna Loa the amount of lag is found to increase with time and is currently about 2yr (figure 1).

## Data Interpretation

The data in figure 4a show that for one certain year, 1988, the maximum of annual variation increases with latitude, the minimum dips down above the equator, however since low $CO_2$ periods are of short duration, the average value continues to increase over the full range of latitude.

The rate of $CO_2$ increase with time over several years is also found to generally increase over the full range of latitude, the maximum being at Point Barrow, one minimum possibly at the Equator and another at the South Pole (figure 5).

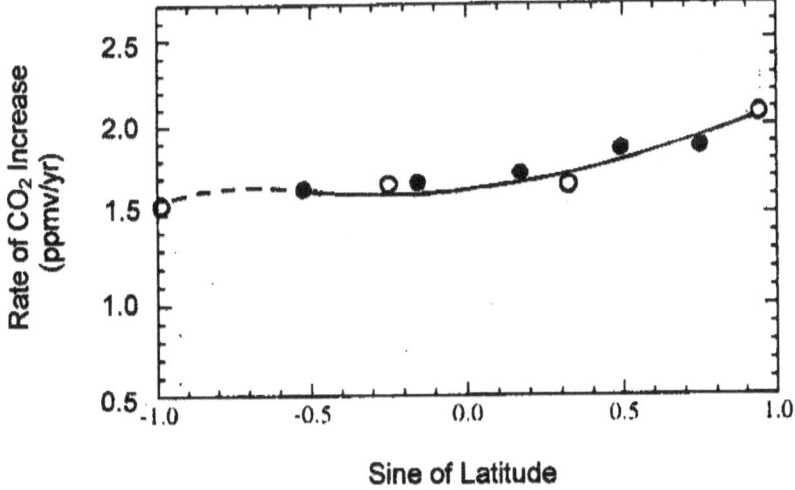

*Figure 5. Average rate of increase of $CO_2$ concentration between 1984 and 1992 as a function of the sine of latitude angle. Slopes were measured at maximum and minimum annual variation and at the estimated average. The three measurements were then averaged.*

Carbon dioxide emitted as a result of burning fossil fuel is found to cause a buildup in the atmosphere. A very large variation in measured interannual contribution occurs for essentially unknown reasons (Keeling et al, 1995), however, a linear average of $CO_2$ growth rate, determined by measurements from all sites, shows that only about 50% of fossil fuel emissions are retained in the atmosphere. The fraction retained over time remains about the same as the rate increases (Figure 6).

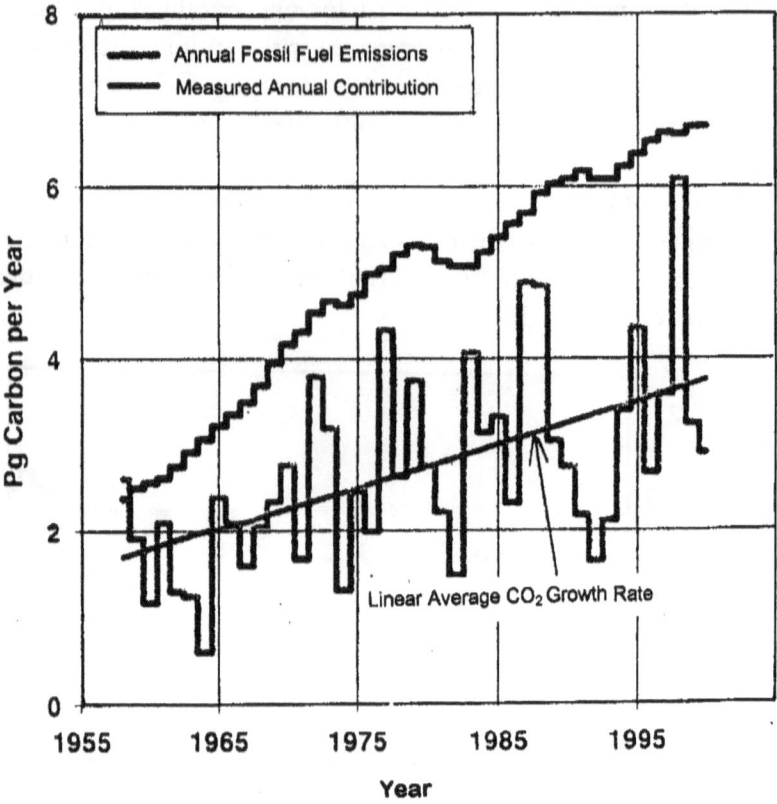

*Figure 6. Estimated annual fossil fuel emissions of CO₂ and the measured annual contribution made to atmospheric CO₂ concentration since 1957. Amounts are expressed in petagrams (billions of tonnes) of carbon (Conway et al, 2003). A linear average of CO₂ concentration shows that the atmosphere retains about 50% of emissions.*

## 3 ANALYSIS

### 3.1 Change in annual variation with latitude

In traditional theory the land biosphere is regarded as the main source and sink for $CO_2$ and, as noted in the introduction, is normally called upon to explain the annual variation of mid-ocean measurements. One of the problems with this is that when the biomass is absorbing $CO_2$, during spring and summer, the processes of burning and decay are somewhat effective at releasing $CO_2$. The two should almost balance each other. If the terrestrial biomass were to be the source there would also be no reason to expect a uniform change in annual variation along one latitude across the whole Pacific Ocean and no reason to expect an increase in variability with increasing latitude especially to as much as 15ppmv. $CO_2$ is a heavy gas and tends to stay in lower levels of the atmosphere and near to the source. This is what makes it possible to have the large variations that are observed. Some mixing, however, does occur and, since that is so, it is unlikely that a low-level concentration profile generated by the biomass, should there be one, would annually be carried out from the land for thousands of kilometers. The variations that are recorded have to be a local effect.

As described above, an aspect of the ocean-as-source-and-sink theory is that $CO_2$ is released from the ocean when the air at the water surface is relatively cool and reabsorbed when it is relatively warm. In the Northern Hemisphere there are very large continental landmasses that get very cold and very hot once a year. This would be expected to cause a variation in air temperature over the ocean and therefore a variation in $CO_2$ concentration. It would also be expected that going north there would be greater variation in temperature and therefore, as observed, greater variation in $CO_2$. In the Southern Hemisphere there are no landmasses that annually get much colder in winter than in summer. This, along with the moderating effect of the relatively large southern ocean, prevents the large air temperature changes and hence the large $CO_2$ variations experienced in the north.

3.2 Annual variability at CMDL sites

3.2.1 Mauna Loa

The average monthly $CO_2$ concentration at Mauna Loa was plotted for the years 1998 and 1990 (figure 7a). Since accumulation is regarded as a rate process, the slope of the concentration curve was measured to obtain a curve showing change of rate (figure 7b). The rate, or first derivative curve, suggests that supply for concentration increase occurs during the winter between October and May and reduction occurs in summer between May and October. Since the rate curve matches with respective expectations for winter and summer it is apparent that it is temperature, and more specifically the relative temperature of atmosphere and ocean, that is the driving force for diffusion across the interface, and that this is the cause of the annual variation. Since the ocean temperature stays relatively constant, $CO_2$ is released when the air is cold and reabsorbed when it is warm.

### 3.2.2 South Pole

At the South Pole an annual variation of about 1ppmv is observed (figure 7c). As with Mauna Loa the first derivative was measured for the years 1989 and 1990 (figure 7d). It is apparent that buildup of $CO_2$ in this case occurs between May and October, southern winter, with maximum rate of supply in June, and that reduction occurs between October and May. It was initially assumed that diffusion would not occur through a solid at Antarctic winter temperatures and that it was a reduction in Brownian movement that caused the observed buildup. But it has to be accepted that it is diffusion out of the ice. The main argument for this is that the maximum rates of change occur at the southern solstices. This would not be expected for a Brownian mechanism because maximum temperature difference would come later, but it would be expected for a thermal diffusion mechanism because the air and sunlight would also affect the ice, reducing temperature difference and causing an earlier peak. It has been suggested by adherents to classical theory that the annual variation at Antarctica is the result of the variation in biomass activity in the Northern Hemisphere. A concentration profile however could not be expected to survive a trip across the southern ocean where there is no profile. It is also claimed, as will be discussed later, that the average reading is less than elsewhere because of a delay in the anthropogenic increase, mainly generated in the Northern Hemisphere, getting that far south. It seems unreasonable to claim that the average concentration is delayed by two years whereas the seasonal variation is transferred within six months, right on cue for southern seasonal change. $CO_2$ variation at the South Pole, whatever the cause, as elsewhere, has to be a local effect. A diurnal flux reversal between air and ice has been observed during a summertime Arctic expedition by researchers from the University of Manitoba (Barber, 2004). This would seem to confirm that thermal diffusion does occur in this circumstance.

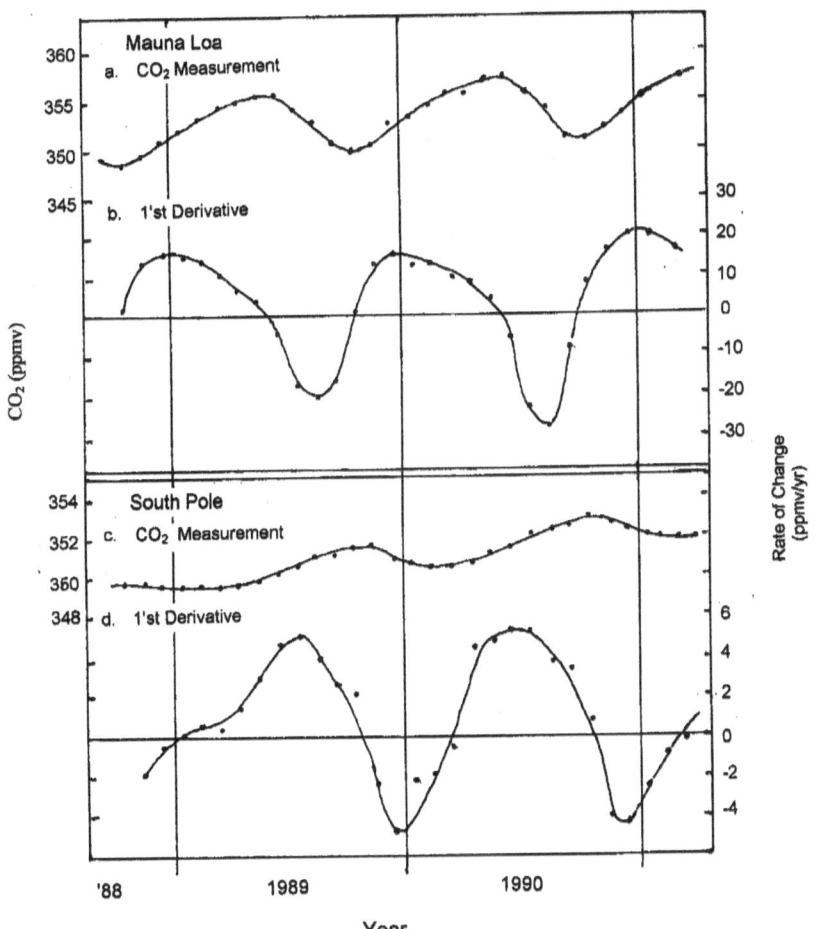

*Figure 7. Measured CO₂ concentration and first derivative variation at the Mauna Loa and the South Pole Observatories for 1989 and 1990. This provides evidence for a physical mechanism for exchange between atmosphere and ocean and between atmosphere and ice.*

### 3.2.3 Samoa

After finding that temperature difference is the source of $CO_2$ cycles in the Northern Hemisphere, the question remains as to why the same thing does not occur at Samoa or in shipboard samples taken south of the Equator. Perhaps, as stated above, it is because of the absence of significant landmass in the south and of the moderating effect of the large southern ocean, both tending to inhibit temperature, and hence $CO_2$ variation.

### 3.2.4 Point Barrow

An attempt was made to treat the Point Barrow data the same way as that of Mauna Loa however it was realized that, without a surrounding ocean, Point Barrow is a special case and more subject to changes in ice coverage than in air temperature. Concentration is fairly steady from December to May, then falls to a minimum in August and September (Figure 8a). Surface albedo measurements published by Larsson et al, (1961)(fig. 8b) show that the albedo within 50 kilometers to the north of Point Barrow also starts to decrease in May and the ocean only becomes ice-free (albedo 20%) during September and perhaps part of August. Partial exposure of water starting in May seems to cause a decrease in $CO_2$ concentration, absorption reaches a maximum during August and September, but then the concentration quickly recovers when the ocean re-freezes. This is discussed further in section 3.4.4.

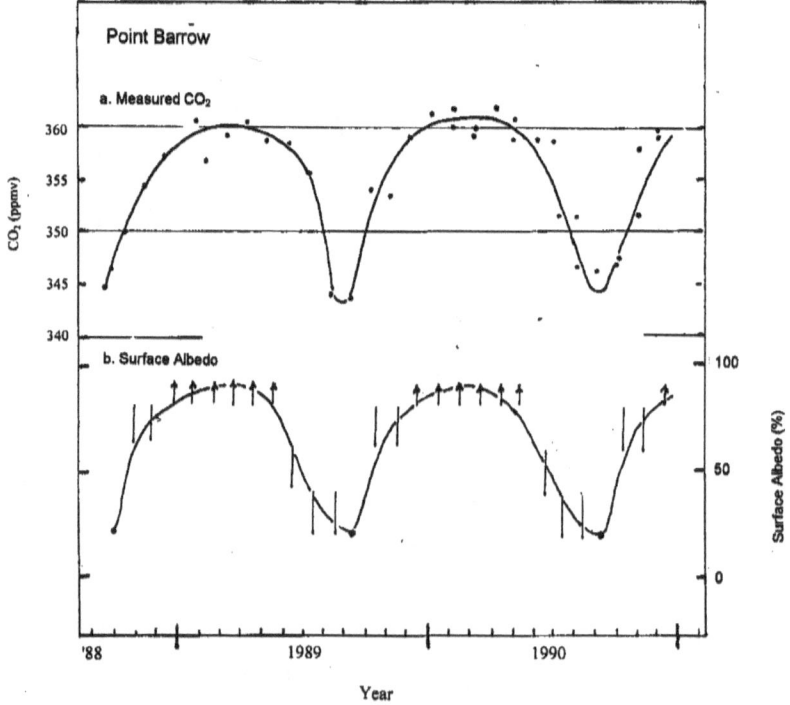

*Figure 8. Graph (a) shows annual variation in $CO_2$ concentration measured at Point Barrow, Alaska. Graph (b) shows an annual variation of surface albedo within 50 kilometers to the north (Larson et al, 1961).*

*The suggestion is that exposure of water causes absorption of $CO_2$ and hence a reduction in concentration.*

## 3.3 Isotope selection

$^{12}C$ isotopes are preferentially released at the ocean surface and there is the same preference for absorption of $CO_2$ by the biomass from the atmosphere. The fact that the $\delta^{13}C$ value for the ocean is mildly positive while that of the atmosphere and, more importantly the biomass, is noticeably negative is an indication, probably something already known, that the bulk of the Earth's compliment of $CO_2$ is in the ocean. Fossil fuel, in the form of coal at least, must also have a $\delta^{13}C$ value similar to that of the current biomass since it is made up of fossilized plants. This is another reason for $\delta^{13}C$ of the ocean to be positive. Since there is a relatively large deposit of coal compared to the volume of biomass, the mildness of the positive value for the ocean is again an indication that the ocean is effectively an infinite source.

During the annual $CO_2$ increase, measured using shipboard sampling, $^{12}C$ is preferentially released with the less preferred $^{13}C$ being left in the ocean. As $CO_2$ increases, the amount of $^{13}C$ in samples, and hence the $\delta^{13}C$ value, decreases. For the decreasing $CO_2$ part of the cycle there is also a preference for $^{12}C$ being the one that actively crosses the boundary, thus $\delta^{13}C$ appears to increase. This is why the curve for $\delta^{13}C$ value closely matches, but changes opposite to, that of $CO_2$ concentration (figure 3). This close matching strongly supports the idea, that diffusion across the air/water interface is what causes the observed $CO_2$ concentration variations and consequently, that the terrestrial biomass is having little, if any, effect on $CO_2$ variation at ocean and other remote sites.

One might ask why $\delta^{13}C$ values are consistently negative and why they decrease with time. The ocean preferentially supplies $^{12}C$ to the atmosphere. That alone would dictate that $\delta^{13}C$ be negative because there would always be less than the standard amount of $^{13}C$ in the atmosphere. The biomass and presumably fossil fuel in the form of coal has a very low $\delta^{13}C$ value. Continuously burning fossil fuel, and coal in particular therefore reduces the relative amount of $^{13}C$ in the atmosphere and causes the decrease in the $\delta^{13}C$ value with time.

An argument is made herein against biological control. Simply put: biological activity should decrease above a certain latitude and limit annual $CO_2$ variation whereas variation continues to increase. Supporting that argument, it is observed that $\delta^{13}C$ decreases with latitude in April (Figure 4), which means that the proportion of $^{12}C$ in samples is increasing as ships travel north soon after the winter season. In conventional theory $CO_2$ change at this time of year would be ascribed to decay of biomass. However if that were the case, the amount of decay and therefore the proportion of $^{12}C$ should start to decrease above a certain latitude, whereas it continues to increase. $^{12}C$ decreases with latitude in August, also as expected in conventional theory, because that is when the biomass is supposed to be absorbing, but again the decrease should not continue to extreme northern climes.

### 3.4 Altitude effects

### 3.4.1 Concentration gradient

An equation derived from Fick's law by Einstein (Jacob, 1999) provides scales for vertical transport, 1-2 days being required for air molecules to migrate upwards 1-3km. This is one reason for altitude effects in $CO_2$ measurements. There may however be another reason. As mentioned above $CO_2$ is heavier than air, in fact each molecule is approximately 44/29 times heavier. Once released at ground level, especially with a continuing supply, a concentration gradient is set up, concentration decreasing with altitude. Concentrations recorded at CMDL sites on a height of land are therefore lower than those obtained by shipboard sampling (figure 4a).

### 3.4.2 Rational for fossil fuel contribution

$CO_2$ emitted by the burning of fossil fuel that doesn't contribute to the buildup in the atmosphere has to go somewhere to meet the requirements of the law of conservation of matter. It is argued in conventional theory that this 50% of extra supply is being absorbed by the biomass. However, rate changes are linear (figure 6), which means that fossil fuel consumption and $CO_2$ buildup increase exponentially with time, and also therefore that the amount to be absorbed increases exponentially. One would not expect the biomass to be able to consume an ever-increasing amount, especially when forests are being cleared for agriculture and wild fires are increasing. The only possibility is that the excess is absorbed by water.

A can of soda water contains a small volume of gas under pressure. A high concentration of $CO_2$ is kept in solution by exchange of $CO_2$ molecules across the gas/liquid interface in the can. When the can is opened the pressure drops and some, but not all, of the $CO_2$ diffuses out. The atmosphere contains about 375ppmv and exchange with this maintains a certain concentration in the water. This is as required to satisfy Henry's law of solubility. When fossil fuel is burned and $CO_2$ is released some of it stays in the atmosphere but some dissolves in the ocean enriching the upper layers to a higher concentration to balance against the higher level in the atmosphere. As it turns out the amount required appears to be about half.

### 3.4.3 South Pole lag

As noted above, the buildup of average $CO_2$ concentration at the South Pole station lags behind that at Mauna Loa. The conventional explanation is that there is a delay of fossil fuel contributions, mainly in the Northern Hemisphere, reaching that far south. This is undoubtedly true and, as described in this report, the northern oceans also make a contribution that takes time to work its way south. The monitoring station on Antarctica is at a high elevation, which would also contribute to lower readings.

### 3.4.4 Northern polar cap

The situation at the North Pole is difficult to explain. Since the dip in $CO_2$ concentration in late summer at Point Barrow station is attributed to ice retreat, the concentration above the ice farther north must be reasonably constant year round. This would mean that the $CO_2$ concentration at the North Pole is higher than at any other place on Earth. In 1989 for example the concentration at the North Pole would have been about 360ppmv, as measured at Point Barrow in winter (figure 8), whereas at Mauna Loa the average was only 352ppmv, at Samoa 350ppmv (figure2), and at the South Pole 349ppmv (figure 1). Because of the distortion of the Earth's spherical shape the North Pole is lower than it is anywhere else. Rates of exchange are low with ice compared to those with water, but still there may be more pressure to diffuse out in winter than to be absorbed in summer, because of a larger temperature difference. A higher local level is therefore required for balance and less is dispersed to the greater atmosphere because of the cold. This allows a buildup and the low elevation and surrounding continental landmass, tends to provide a container of sorts to maintain it.

3.5 Rational for global $CO_2$ buildup

3.51 Change in buildup with latitude

As noted above, the rate of $CO_2$ concentration change with time increases with latitude (figure 5). A mechanism was devised to allow the atmosphere above the ice of the Arctic Ocean to have higher than normal $CO_2$. It is not likely that excess low-level $CO_2$ would spill out from the arctic basin to cause readings to be higher south of Alaska, and to have the effect spread south as far as 18° N. The effects have to be locally induced. As we have postulated, cold air coming from Siberia in the northern winter causes $CO_2$ to be released from the ocean. In summer, hot air from the land causes $CO_2$ to be absorbed. Since temperature variation increases with latitude it is also expected, as observed, that $CO_2$ variation would increase in the same way. Perhaps even though some $CO_2$ released in winter remains in proximity, there is insufficient time for the same amount to be reabsorbed in the short summer. This may be one reason there is the observed buildup, increasing with latitude. Clearer skies cause hotter summers and colder winters but the resulting release and absorption may not balance each other. In summer both the air and water surface are heated without being inhibited in any way, thus decreasing temperature difference and limiting absorption, whereas in winter the water surface can only be cooled so much because extra energy has to be lost to allow freezing. This provides greater temperature difference in winter than in summer and thus more release than absorption. This is another reason why there is the trend that is observed. The northern Pacific Ocean and presumably the northern Atlantic as well, is acting effectively as a pump, supplying $CO_2$ to the environment. The difficulty is to assess the global implications. Three scenarios are proposed but perhaps the answer should include some of all three.

### 3.5.2 Scenario 1, Access

Once fossil fuel emissions are partially mixed into the greater atmosphere they are not effectively removed either by diffusion at the ocean surface or by rain or other weather effects. For release of $CO_2$ there is always a ready supply in the surface water, but for absorption there is a less reliable supply because the $CO_2$ has to find its way back to the surface to be reabsorbed. This maintains the buildup without any contribution from northern oceans.

### 3.5.3 Scenario 2, Absorption offset

$CO_2$ readily dissolves in water and in order to support the increasing level of concentration in the atmosphere it is required to have the cloud cover adjust itself to maintain that level. As $CO_2$ increases skies get clearer in northern winter and more is released. With this scenario the higher concentration above the northern ocean is that required to be distributed to the global atmosphere to compensate for absorption and thus maintain the buildup.

### 3.5.4 Scenario 3, Exponential release

The higher concentration above the northern ocean is the measurement of a supply in excess of that provided by fossil fuel emission. According to the developing theory, clearer skies allow the $CO_2$ concentration to increase. It could also be argued therefore that the increase in concentration in turn provides clearer skies, which means that there is positive feedback and a possibility of exponential growth. It is observed that this does indeed happen at the beginning of interglacial periods (Bell, 2003). During a glacial stage the ocean is warmer than the atmosphere because of temperature lag, which means that it is relatively cloudy and the climate keeps getting colder. Eventually the ocean begins to freeze causing its temperature to drop below that of the atmosphere, the skies start to become clear, and $CO_2$ is released from the ocean. This signals the end of the glacial stage: the more $CO_2$ that is released, the greater the disparity between atmosphere and ocean temperatures, the clearer the skies become and the more rapid the release. Equations for release have been determined in as yet unpublished work (Bell, 2005). It starts out being exponential with time but eventually reverts to a linear rate of approximately 10ppmv/1000yr. With present disparity, global release rate should be in the linear range, i.e. 0.01ppmv/yr. Perhaps the increase in measured rate of release in the northern Pacific Ocean observed herein, about 0.2ppmv/yr, is this extra disparity-induced contribution, exaggerated because the $CO_2$ hasn't had time to disperse and the volume of air containing the contribution is small in comparison to that of the whole atmosphere.

## 4 DISCUSSION

Available $CO_2$ data supports the proposed theory of a diffusion mechanism for release and absorption by and from the ocean. There is much to support the idea, however the main evidence for it is the finding that the variation in $CO_2$ concentration increases with increasing latitude over the northern Pacific Ocean and that the rate of release peaks at the coldest part of northern winter. A physical process, as proposed, could explain this result whereas the chemical and biological processes of conventional theory could not, because their influence would decrease with increasing latitude. The $\delta^{13}C$ data also supports the idea since observed trends for this value can only be explained by a diffusion mechanism. The $\delta^{13}C$ value cycles opposite to $CO_2$ concentration during annual cycles, thus the two must be related and both therefore must involve a diffusion mechanism.

The proposed diffusion theory of climate change requires that skies become clearer as atmospheric $CO_2$ concentration increases. In Bell (2003), a concept called a cloud cover factor was introduced to explain events in the early stages of an interglacial period. When skies are deemed to be clear, at the start of the period, the factor contributes to atmosphere temperature above that associated with increasing $CO_2$. When skies eventually become cloudy the factor detracts from atmosphere temperature. At the time of publishing Bell (2003) the factor could not be identified. The thought has occurred during the current analysis that the cloud cover factor is related to the Earth's albedo. When atmosphere $CO_2$ concentration increases there is an associated increase in temperature due to greenhouse warming, but because of the related decrease in cloud cover there is also a reduction in albedo, which adds to the temperature increase. When cloudy conditions become prevalent the opposite happens. The cloud cover factor is thus a loss, or gain, of the Earth's albedo. This provides more empirical evidence to support the proposed theory.

*Data Interpretation*

Over the longer history of the Earth cloud cover was automatically adjusted so that there was a balance between release and absorption of $CO_2$. When climate control oscillation began at the beginning of the Pleistocene Epoch simple cloud control was lost and the $CO_2$ concentration has cycled up and down at the whim of opposing ocean and atmosphere temperatures, which lag each other by thousands of years (Bell, 2002 & 2003). $CO_2$ comes out of the ocean when the *atmosphere* is relatively warm and the climate relatively *clear* and is reabsorbed when the *ocean* is warm and the climate *cloudy*. This is well demonstrated to be true by empirical evidence.

There is a problem rationalizing this with the findings of this report. The theory just described requires buildup of $CO_2$ to occur when skies are clear and the atmosphere is *warmer* than the ocean. On examination of the data however, it has to be admitted that release happens when the atmosphere, or at least the air in contact, is *colder* than the ocean. Is it possible for both to be true? For clearing skies the original idea was that a cooler ocean would allow less evaporation and a warm atmosphere would be less likely to form clouds. This still seems to be true. Clearing skies do warm most of the planet, aided and abetted by increasing $CO_2$, but it is not the warmth and not the sunlight, nor is it a change in insolation from a change in the Earth's attitude with respect to the Sun that releases the $CO_2$, but rather, as we have seen, it is the cold of long winters in the far north, which is enhanced by clear skies.

5 CONCLUSIONS

5.1 Air temperature determines the direction of $CO_2$ flux across the air/water interface of the ocean, relatively cold air causing release and warm air, absorption.

5.2 Large continental landmass in the north causes an annual variation in temperature of air spreading over the northern Pacific Ocean and that in turn causes a variation in local $CO_2$ concentration. Greater air temperature variation with increasing latitude causes greater $CO_2$ variation.

5.3 In the Southern Hemisphere there is no landmass subjected to the same temperature extremes hence no measurable annual variation in $CO_2$ concentration.

5.4 $\delta^{13}C$ values in ocean-site samples follow but oppose $CO_2$ concentration, and variation in this case is thought to be the result of a preference for diffusion of $^{12}C$ over $^{13}C$ in both directions across the water surface. The idea that interface diffusion is the important mechanism for $CO_2$ change is strongly supported by $\delta^{13}C$ data.

5.5 Because $CO_2$ is heavier than air, its concentration decreases with elevation when there is continuing release at ground or water-surface level.

5.6 It is proposed that, about half of the $CO_2$ generated by human activity, goes into the atmosphere while the other half is absorbed by upper layers of the ocean to balance diffusion across the interface.

5.7 Antarctic $CO_2$ concentration may lag that measured at other sites partly because of an altitude effect. The annual variation at this monitoring station is tentatively related to diffusion into the ice.

5.8 Annual $CO_2$ variation at Point Barrow is caused by an annual variation in the amount of sea ice to the north of the station. The $CO_2$ concentration above the Arctic ice is expected to be higher than it is anywhere else because of unbalanced diffusion from the ice and because of the low elevation.

5.9 From annual variation of $CO_2$ and $\delta^{13}C$ it may be determined that, contrary to conventional theory, absorption and release of $CO_2$ by the terrestrial biomass has no effect on remote site measurements.

5.10 Annual variation and the rate of increase of local $CO_2$ concentration in the Pacific Ocean both increase with increasing latitude. There is a suggestion from this that northern skies are gradually clearing, winters are becoming colder and as a result, oceans are supporting, or actually contributing to, global $CO_2$ buildup.

5.11 The theory, which proposes that the ocean is the major source and sink for $CO_2$, may be rationalized to suit most observations and does this more effectively than traditional theories of climatology.

Acknowledgements

The author appreciates continuing advice and support by John C. Anderson. Thanks are expressed to Pieter Tans and Tom Conway of the Climate Monitoring and Diagnostics Laboratory, NOAA, for permission to copy figure 6 and to use web site data from $CO_2$ monitoring stations. Permission to reproduce figures 2, 3 and 4, previously published in the Journal of Geophysical Research, (Morimoto et al, 2000), has generously been granted by the American Geophysical Union.

References

Barber D. 2004 University of Manitoba, Private communication.

Bell LG. 2002. World Ocean Temperature Lag Time: An Analysis Based on Glaciation Data for the Last Two Million Years. *Theoretical and Applied Climatology.* **73 (3-4)**: 243-247.

Bell LG. 2003. Ice Age Mystery: A Proposed Theory for the Cause of Long Term Climate Change. *Theoretical and Applied Climatology.* **74 (3-4)**. 245-253.

Bell LG. 2005. Ocean/Atmosphere Temperature Disparity Effects: An Analysis of CO2 and Temperature Data for the Previous Interglacial Period. (In preparation).

Borg RJ, Dienes GJ. An Introduction to Solid State Diffusion, 3. Thermomigration-Diffusion in a Thermal Gradient. 145-148, Academic Press: Toronto.

Conway T. and Tans P. 2003. *Climate Monitoring and Diagnostics Laboratory,* NOAA ftp://ftp.cmdl.noaa.gov/ccg/co2/in-situ/ for data and http://www.cmdl.noaa.gov/ publications/annrpt26/index.html for figure 6.

Higuchi K, Nakazawa T. 2001. Carbon Dioxide, Recent Atmospheric Trends, *in Encyclopedia of Global Environmental Change.* Munn RE ed. **1**: 254-260, John Wiley and Sons: Chichester.

Jacob DJ. 1999. *Introduction to Atmospheric Chemistry.* Princeton University Press: Princeton,

Keeling CD. 2001. Keeling, Charles David. *Encyclopedia of Global Environmental Change,* Munn RE ed. **1**: 484-485. John Wiley and Sons: Chichester.

Keeling CD, Whorf TP, Wahlen M, and van der Pilcht J. Interannual Extremes in the Rate of Rise of Atmospheric Carbon Dioxide Since 1980. *Nature*, **375**, 666-670, 1995.

Larson P, Orvig S. 1961. Atlas of Mean Monthly Albedo of Arctic Surfaces, Meteorology No. 45, *Arctic Meteorology Group*, McGill University: Montreal.

Masterton, W.L., E. J. Slowinski (eds). 1997. *Chemical Principles*. W. B. Saunders Co.: Philadelphia: 292.

Morimoto S, Nakazawa T, Higuchi K, Aoki S. 2000. Latitudinal Distribution of Atmospheric $CO_2$ Sources and Sinks Inferred by $\delta^{13}C$ Measurements from 1985 to 1991, *Journal of Geophysical Research*. 105, NO. D19: 24315-24326.

The basic theory being developed; namely that the ocean is the main source and sink for $CO_2$ and exchange between atmosphere and ocean is the result of thermal diffusion, etc., is supported by this analysis of actual recent data. The $\delta^{13}C$ data from shipboard sampling is the most compelling in this regard. There is also, interestingly, an indication that thermal diffusion of $CO_2$ even occurs between air and ice.

Of major significance, as far as future predictions are concerned, is the conclusion that a portion of introduced $CO_2$, from whatever source, is automatically retained in the atmosphere by the current layout of land and ocean.

# Chapter 7
## Orbit Evidence

As mentioned in earlier chapters, one of the pillars of mainstream climatology is the idea that the Earth's orbit eccentricity cycle, acting in concert with precession and possibly other attitude cycles, has caused the glaciation cycles of recent history. A study, as reported in this chapter, was undertaken using ideas of the developing new theory to determine exactly what influence these cycles have. Andre Berger is the recognized authority on the subject. Many of his reports include the figure, attributed to Berger and Loutre, shown here as Figure 0-2. This is supposed to provide evidence that the three cycles of Earth motion interact to produce intermittent glaciation. No explanation is offered as to how this interaction is supposed to occur.

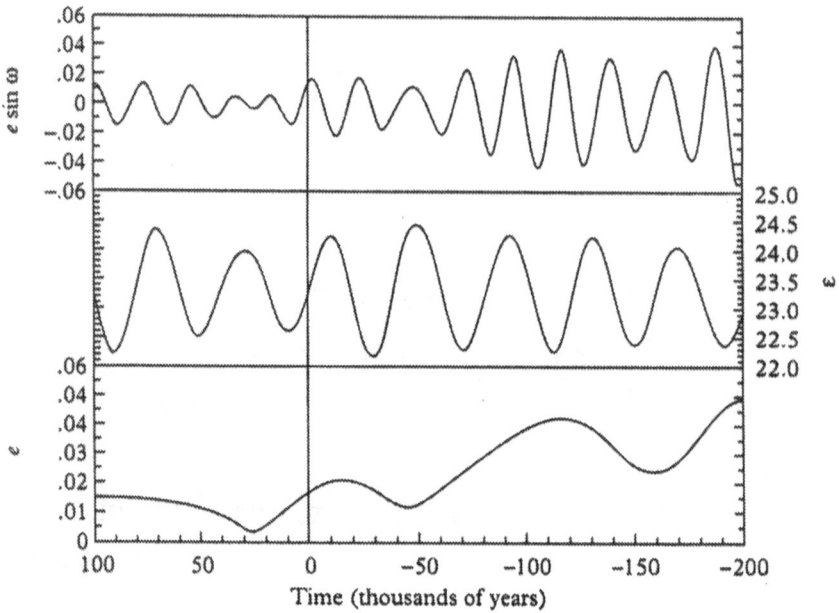

*Figure 0-2. Long term variations of precession, obliquity and orbit ellipse eccentricity for the past 200,000yr and future 100,000yr. (From Berger and Loutre, 1991). The ω curve is supposedly an envelope of possible obliquity from precession and 41,000yr nutation.*

The report of this chapter was an effort to find long-term augmenting interaction between winter and summer precession cycles, with periods of 22,000yr and 30,000yr respectively, as well as the much weaker nutation cycle of 41,000yr period, with the eccentricity at 100,000yr. This was unsuccessful, however one cycle was found that could indeed interact with orbit eccentricity. The interaction affects glaciation but does not determine the period of glaciation cycles. This is the main thesis of this chapter and follows in sequence those of chapters 1 and 3.

Note: As evidence for obliquity from precession being currently at maximum, information from Berger and Loutre was referenced and figure 0-2 was adapted and included as figure 1 of the report pasted below. Part of the original figure 0-2 was edited out because it is claimed by the authors to be an *envelope* of obliquity change for precession and nutation. An envelope should have upper and lower bounds and to include the figure as published would be to accept a very exaggerated influence of the relatively weak nutation cycle.

# 7. Evidence of an Effect of the Earth Orbit Cycle on Glaciation

## L. G. Bell

*Abstract*

*The Earth's axis is known to tilt toward and away from the Sun in time with precession. This produces alternating winter and summer season maxima. It is found that at every 15.5 revolutions of the Earth's precession, equivalent to 4 orbit eccentricity cycles, alternating northern season maxima occur at the aphelion of the orbit ellipse at maximum eccentricity. The opposing season is then said to be in sync with the perihelion. It is theorized that when winter is in sync there might be slightly enhanced glaciation and when summer is in sync slightly reduced glaciation. Data is found to support the idea to some extent as far back as 1.6myr with enhanced and reduced glaciation alternately occurring at 400,000yr intervals. The finding that combined precession and orbit effects last such a long time leads to the conclusion that the two Earth-attitude cycles are associated. Evidence of enhanced glaciation, with a period of 800,000yr, rather than 100,000yr, also provides more assurance that effects other than the period of the orbit cycle have influenced the period of recent glacial/interglacial cycles. A confluence of many astronomical and earth temperature-nodal effects occurred at 800,000yr BP and probably prompted buildup of permanent ice sheets on Antarctica and Greenland. Support is provided for the previously developed idea that ocean/atmosphere temperature disparity and its effect on cloud cover along with biomass intervention is what has determined recent climate cycling.*

## 1 INTRODUCTION

The Earth orbit is an ellipse with the Sun at one focal point and eccentricity going through a maximum and minimum with a period of about 100,000yr. By itself this eccentricity cycle, sometimes referred to as the orbit cycle, has not been expected to cause a variation in heat input sufficient to affect glaciation. It has been assumed however, since the idea was proposed by James Croll in the year 1875 (Ekman, 1993), that it acts in concert with other Earth-attitude cycles to produce the approximate 100,000yr period of recent major glaciation events. Calculations by Milankovic in the 1930s, of variation in induced latitudinal insolation, increased the credibility of this idea and gave impetus to the general theory of astronomical climatology. In the light of recent developments it is not at all obvious that these long-held beliefs have merit. An historical record of the extent of global glaciation rates has recently come available. The objective of the present study is to examine what influence the motion of the Earth has had on the glaciation record and how gravitational forces of other members of the Solar System, and other Earth-bound cycles affect it. (The large southern ocean has a moderating effect on the climate of the Southern Hemisphere. Because of this the north is where most major glaciation and other climate events occur and since they dominate, climates referred to herein are those of the Northern Hemisphere, unless otherwise stated).

In an earlier report (Bell, 2002) it was found, using oxygen-ratio glaciation data obtained from ocean sediment cores (Porter, 1989), that the 22,000yr (winter) precession climate cycle and a 23,500yr ocean/atmosphere temperature cycle form harmonic nodes of enhanced glaciation at 352,000yr intervals. The two cycles as described were also found to interact with the weaker 41,000yr-nutation cycle to form grand nodes with period of 2.46myr. A grand node occurred at 800,000yr BP, with enhanced glaciation effects resulting from it centered at 620,000yr BP. Evidence of the mid-point of the grand harmonization is found at 1.85myr BP, the difference between these two major glaciations being 1.23myr. There was no suggestion in the study that the orbit cycle has had an influence on glaciation.

In another report (Bell, 2003b), again using oxygen-ratio glaciation data, evidence was found for a 30,000yr precession-related climate cycle. It was believed to be a summer cycle since it was observed to have a negative effect on glaciation. Attempts were made to find harmonic nodes with the 30,000yr cycle and other climate and temperature cycles as well as the orbit cycle, but without success. It was suggested in this second report that the precession cycle itself, with 25,800yr period, may act in concert with the orbit cycle. It was also suggested that the greatest effect would be observed when one of the precession obliquity maxima, summer or winter, was *in phase* with the orbit perihelion at maximum eccentricity. It is now realized, as will be shown in the present report, that summer and winter maxima can only be in phase with the ellipse aphelion. (It is possible for seasons to be in phase with perihelion in the Southern Hemisphere because precession maxima occur at the same time as in the north but on the other side of the Sun and therefore at perihelion. Orbit and precession effects add to each other in the south whereas in the north they subtract. The moderating effect of the southern ocean rules out any resulting significant effect on glaciation).

With this development it is realized that the interaction of precession and orbit cycles is not as spectacular as it could have been had alignments been different. It is considered worthwhile, however, to go through the exercise of describing the interaction that does occur and to examine what evidence there is for it. This is deemed to be important because it affects how we think about what causes climate change.

## 2 INFORMATION

### 2.1 Disparity effects

This report is one of a series describing and developing a new theory on the cause of climate change. It is maintained in Bell, (2003a), and other as yet unpublished work, that ocean/atmosphere temperature disparity affects the amount of cloud cover and that in turn determines the release and absorption of $CO_2$ by the ocean. This information is used in the present work to aid in interpretation of glaciation effects.

### 2.2 Orbit characteristics

As mentioned above the Earth orbit is an ellipse. The eccentricity $e$, defined as $(a^2-b^2)^{1/2}/a$, where a is the semi-major axis and b is the semi-minor axis, never reaches 0.00 but may range as high as 0.07 and varies about 0.03 for any particular orbit cycle. Calculations by Berger, (1977) (Figure 1) show that we are presently in the ellipse phase with maximum eccentricity having occurred about 15,000yr BP. The calculations confirm that maxima of ellipse eccentricity should be expected at approximate 100,000yr intervals.

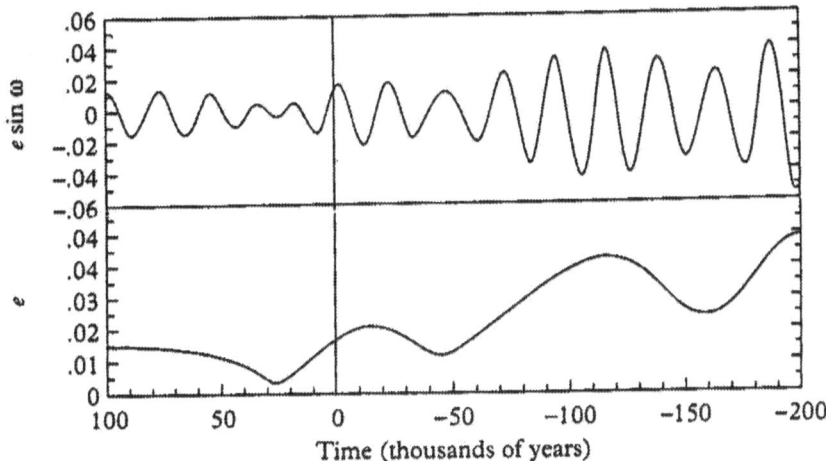

*Figure 1. Long term variations of precession contribution to insolation, e sin ω, where ω represents obliquity and e orbit ellipse eccentricity, for the past 200,000yr and future 100,000yr. (From Berger and Loutre, 1991). Of particular interest is the present-day maximum in e sine ω indicating summer precession maximum at aphelion.*

## 2.3 Precession component

Precession, which requires that the seasons precess around the Earth's orbit, has a period of about 25,800yr (Bishop, 2001). Looking down at the Solar System from the north, seasons would be observed to turn clockwise, because of precession, while the Earth revolves counter clockwise. Perihelion of the orbit ellipse is now occurring on January 2. The winter solstice has therefore passed through the perihelion about 800 years ago, a very short period of time as far as Earth attitude cycles are concerned. Going back 26,600yr BP, slightly more than one precession circuit, winter precession minimum was about equal distance on the other side of maximum orbit eccentricity.

The average obliquity, or angle between the plane of the equator and the ecliptic, is about 23.4°. Obliquity is not constant and is reported to change from about 21.8° to 24.4°. The greater part of this variation is attributed to precession, which is caused by gravity effects of the Moon and Sun, and occurs with the same period as precession. That is to say the Earth axis tilts toward the Sun to produce the northern summer maximum insolation at intervals of 25,800yr. Meanwhile northern winter, being opposite, is going through minimum insolation at the same times. Every 12,900yr the roles reverse. Another, relatively minor, part of obliquity change, commonly referred to as the 41,000yr nutation cycle, is caused by an interaction of the Earth's orbit with the orbits of other planets, mainly Jupiter and Venus.

Calculations of change in eccentricity and insolation over time have been made by A. Berger and published in various journals, e.g. as that shown in figure 1. From figure 1 it is possible to determine that summer on the precession circuit is currently, 1: at maximum and 2: at the ellipse aphelion. The reasoning is that 1: increased insolation from precession is sufficient to overcome the reduction imposed by being at the orbit aphelion, to thus produce a current maximum in the value of $e$ sine $\omega$, and 2: it is opposite winter, which is at perihelion. Winter, it may also be concluded, is at precession minimum.

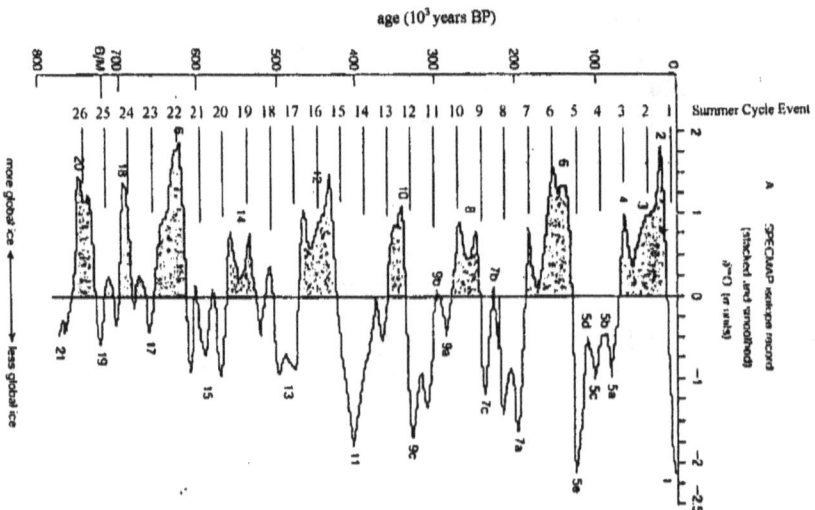

*Figure 2. Marine oxygen isotope records obtained by Porter (1989), Positive $^{18}O$ to $^{16}O$ ratio readings in ocean sediments produce shaded areas interpreted as glacial stages. Of interest to the present study is the approximate 100,000yr periodicity of recent glaciation, the pronounced deglaciation profile during the interglacial period, 400,000yr BP, and the decreasing negative effect of indicated summer cycles on the glaciation rate since that time.*

## 2.4 Glaciation data

Figures 2&3 provide oxygen isotope ratio data from ocean sediment cores (Porter, 1989) interpreted as rate of glaciation (Bell, 2003b). Shaded areas above the horizontal line are periods of increasing glaciation and areas below the line are periods of decreasing glaciation. The two figures are from different sets of samples, grouped together based on core sample availability, such that one set of data gives a longer record than the other. Figure 2 containing the shorter-term data, is as used in Bell, (2003b) and has 30,000yr summer precession cycles indicated.

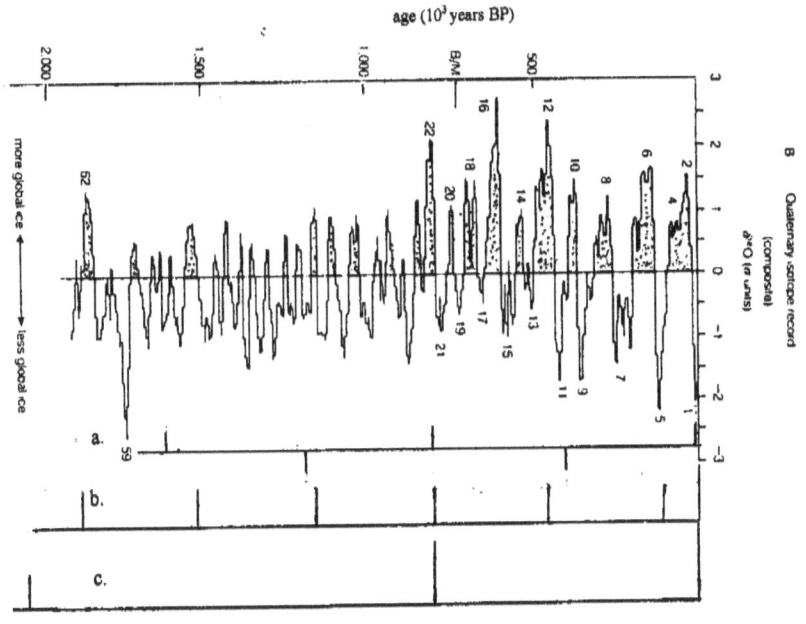

a. season in- sink events

b. 352,000yr harmonic nodes

c. 2.46myr grand harmonic nodes

*Figure 3. Long term glaciation rate data of Porter (1989). The timing of expected positive and negative effects on glaciation, from winter and summer seasons in sync with orbit ellipse perihelion, from 352,000yr harmonic nodes and also from 2.46myr grand harmonic nodes (Bell, 2002), are indicated. The glaciation effect of all three acting together at 800,000yr BP is centered at 620,000yr BP (peak 16) and that of the mid-grand node at 1.85myr BP (peak 62).*

This was included because it was thought that the summer cycles might help identify any orbit-related glaciation effects. Figure 3 contains the longer-term oxygen ratio data as used in Bell, (2002) and shows the location of nodes and grand nodes described in the introduction. The node information is also expected to aid in identification of glaciation anomalies. Figure 3 also shows, incidentally, that the period of glacial/interglacial cycles increases with time.

## 2.5 Glaciation observations

It will be noticed in figure 3 that the period of glaciation cycles only generally increases with time. The period depends to some extent on glaciation rate amplitude, which does not increase uniformly. Occasional periods of enhanced glaciation may be observed, at the nodes mentioned and especially during a 400,000yr period between about 800,000yr and 400,000yr BP, where the average rate of glaciation stays well above zero. The amount of glaciation occurring during glacial stages of this 400,000yr period is not matched by melting during brief intervals of deglaciation. It may also be observed that the duration of subsequent interglacial/glacial cycles, the last four before present, increased from about 90,000yr to 125,000yr, most of the increase occurring in the glacial stage.

## 3. ANALYSIS

### 3.1 Cycle relationships

It was theorized in Bell (2003b) that the orbit cycle can only have an effect if one of the major seasons is in phase; i.e., when precession maximum occurs at the perihelion of the orbit cycle at maximum eccentricity. It is recognized now that this is not possible.

To explain, it is necessary to adopt new terminology. Let us say that when a season at precession *minimum* is at perihelion of maximum ellipse, it is *in sync*. It was reasoned above that winter precession minimum is currently at perihelion and is therefore in sync. Going back in time 4 precession cycles 4 x 25,800yr = 103,200yr, winter would have been back to the perihelion, but 3,200yr out of sync. In 31 precession cycles 31 x 25,800yr = 7,998,000yr, or 8 orbit cycles, winter would have made the complete circuit and be back to where it is now, once again in sync. Since summer change is opposite, we may conclude that in 15.5 precession cycles, 15.5 x 25,800yr = 399,900yr, or 4 orbit cycles, summer would have been in sync and 4 orbit cycles before that, 31 precession cycles before present, it was winter in sync and 4 orbit cycles before that it was summer, etc.

When winter is in sync, as it is now, winter at perihelion is colder (weaker) because it is at precession minimum but warmer because it is closer to the Sun. Summer on the other hand is stronger from precession but farther away. It is just possible for some glaciation enhancement in this circumstance because winter and summer have a chance of being cooler for different reasons. In four orbit cycles, roles are reversed with winter stronger but farther away and summer weaker but closer to the Sun, which might produce warmer conditions and hence a reduction in glaciation. If this were to be so we might expect glaciation rate to have been reduced at 400,000 and 1,200,000yr BP, when summer was in sync and enhanced at 800,000 and 1,600,000yr BP, when winter was in sync. There might also be slight tendency for glaciation buildup, or reduction, during adjacent eccentricity maxima since the seasons do not go far out of sync in the intervening 100,000yr.

3.2 Glaciation evidence

The tendencies for in-sync glaciation effects can be overshadowed, or aided and abetted, by other things that are happening, such as glacial/interglacial cycles, harmonic nodes and harmonic grand nodes; the latter two with periods of 352,000yr and 2.46myr respectively (Bell, 2002).

Of interest in figure 2 is the pronounced drop in oxygen ratio during the interglacial period at 400,000yr BP indicating rapid deglaciation, the negative influence we are looking for. There also appears to be, as one might expect, a reduction of the negative influence of the summer cycle on glaciation since that time. Figure 3 shows a longer-term record with glaciation nodes marked at 352,000yr intervals. The expected positive and negative points of influence by seasons in sync with the orbit cycle are also marked. Again, as in figure 2, there is the pronounced interglacial oxygen-ratio profile indicating pronounced deglaciation at 400,000yr BP following a very severe period of glaciation, when it should not otherwise have been expected. Then there is enhanced glaciation at 800,000yr BP; winter in sync combining forces with an harmonic node. At 1.2myr BP, summer in sync almost cancels the adjacent harmonic node, and finally at 1.6myr BP, winter in sync again combines forces with a nearby node to produce enhanced glaciation. A pictorial version of orbit arrangements for season in-sync events and related glaciation trend is shown in figure 4.

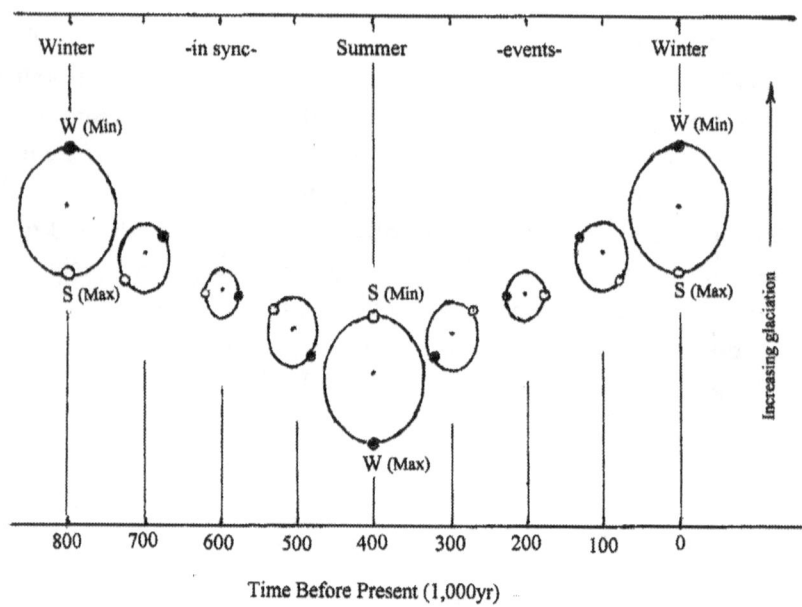

*Figure 4. A pictorial version of alternating winter and summer season precession extremes for each orbit cycle, beginning 800,000yr BP. In sync events and the expected effect on glaciation are indicated.*

### 3.3 Super node

800,000yr BP is a particularly interesting time. As described above it is the timing of one of the 352,000yr-period harmonic nodes formed by the 22,000yr-precession climate cycle and a 23,500yr-ocean/atmosphere-temperature cycle. It is also the timing of a grand node between the two cycles just mentioned and the planet-induced, 41,000yr-nutation cycle. Now we find that an in-sync reaction between winter precession and orbit eccentricity also occurs at 800,000yr BP. The confluence of all these glaciation-enhancing effects produces a period of intense glaciation at 800,000yr BP as shown in figure 3. Glaciation tends to feed on itself thus the full impact of glaciation triggering events may be delayed. The maximum rate of glaciation thus occurred at about 620,000yr BP and the climate was adversely affected for 400,000yr. Since oxygen-ratio glaciation data is a rate, the area under the curve above zero represents ice volume being accumulated; area below zero is the amount melting. Over the history, for which there is a record, there is no other period of imbalance between glaciation and deglaciation, thus the buildup during this time of high glaciation rate probably represents the volume of ice stored on Antarctica and Greenland. The severely cold climate only ended with the advent of summer in sync and coincident enhanced summer cycle events around 400,000yr BP. The relative volume of ice stored, by the way, during the 400,000yr period between 800,000 and 400,000yr BP and throughout history, could be calculated by integration of the glaciation rate as was done for the last two glacial/interglacial cycles in Bell, (2003b).

3.4 Recent glaciation periodicity

As noted above, the average duration of the last four glacial/interglacial cycles is about 100,000yr and as further noted, the period has actually been increasing rather dramatically due mainly to an increase in the duration of glacial stages. It seems incumbent upon us to explain why.

It was theorized in Bell, (2003a) that the temperature of the ocean relative to that of the atmosphere determines whether the climate is clear or cloudy. It was proposed that the Ice Age, or last glacial stage, ended because the ocean began to freeze, thus allowing it to cool faster than the atmosphere and thereby, eventually to provide the clear skies that ended the glacial regime. The peak in oxygen ratio toward the end of the glacial stage in figure 2, is evidence for this ocean freezing. The second glacial stage before present seems to have ended the same way, however the reading for an ocean-freezing peak in this instance was affected by a summer cycle event. Perhaps the event prevented freezing above test sites. The third and fourth glacial stages before present did not exhibit oxygen ratio peaks and likely therefore did not end by the ocean freezing. It appears rather that they were ended by summer cycles (figure 2).

According to the developing theory, all that is required to end a glacial stage is to have the ocean become cooler than the atmosphere. This causes a change from cloudy to clear conditions, which can be achieved by either cooling the ocean as in stages one and two before present or by heating the atmosphere as in stages three and four. The change that makes this possible has to do with in-sync astronomical events. (As will be discussed later it is also possible to have a glacial stage

135

end by an increase in $CO_2$ early in the glacial stage). Summer in sync occurred at 400,000yr BP, thereby increasing atmosphere temperature as well as the strength of summer cycles. This would not greatly affect the duration of the interglacial periods, however, because of the relatively modest glaciation rate, summer cycles were apparently strong enough to provide the temperature crossover required to end glacial stages 4 and 3. For stages 2 and 1, as winter in sync approached, summer cycles were not strong enough to effect the change and it became necessary for the ocean to freeze, which takes much longer.

## 4. DISCUSSION

Indications are that glaciation data are meeting expectations of enhancement by summer and winter seasons predicted to have been in sync. This confirmation of the theoretical interaction of orbit and precession cycles adds more weight to the argument put forth in the present series of reports that the orbit cycle does not determine the period of glaciation cycles. Enhanced climate effects are mild and occur at 800,000yr intervals, not 100,000yr. It was found earlier (Bell, 2003a) that the timing of glacial periods is determined by a major temperature cycle of the ocean. It is just a coincidence that during the last few hundred thousand years the average timing of glacial cycles is about the same as that of the orbit cycle.

Again, according to the proposed theory, the climate in the distant past was controlled by the amount of cloud cover. Should there have been, for example a temperature increase, there would be more evaporation and more clouds. This would cause the Earth's albedo to increase, which would reduce the temperature and thus maintain control. This was fine as long as the ocean temperature was unaffected. At the start of the Pleistocene Epoch, as Antarctica and eventually also Greenland, moved to their present positions, climate change was such as to change the ocean temperature. Cloud cover then worked against control and, with accommodation now provided for extensive glaciation, the system effectively went into oscillation.

Climate cycles during the Pleistocene Epoch have not been increasing steadily, as expected for a control system exhibiting oscillation. This is because cycle size is dependent on more than one set of variables. The expected gradual increase in glaciation magnitude and cycle period has superimposed upon it the effects of the interactions of extra-terrestrial and earthbound cycles as described above. Glaciation lag is another factor to be considered. In addition a release and absorption of $CO_2$ occurs during each major transition. All of these factors affect the rate of glaciation and hence the period of glacial/interglacial cycles.

In spite of many competing factors, some generalizations can be made. At the start of an interglacial period skies are becoming relatively clear. With diffusion across the air/water interface being from warm to cool, $CO_2$ is released from the ocean in cooler parts of the planet (Bell, 2005b), average surface and atmosphere temperatures increase and the ice starts to melt. As ocean/atmosphere temperature disparity increases skies get clearer, northern winters get colder and the rate of release increases. Increasing biomass eventually stops the increase of $CO_2$ and consequently atmosphere temperature, thus allowing the ocean temperature to eventually catch up and exceed that of the atmosphere. The skies then become cloudy and $CO_2$ is gradually reabsorbed until the atmosphere temperature falls low enough to allow glaciation to begin. Climate conditions at which biomass intervention occurs and also at which the onset of glaciation occurs are always about the same and thus the duration of interglacial periods stays fairly constant.

The duration of glacial stages, on the other hand, exhibits greater variation because it depends on the ocean temperature being allowed to fall below that of the atmosphere. At the start of a glacial stage the atmosphere temperature drops drastically because heat is reflected away by the ice. The spread of glacial ice causes the northern forests to retreat and decay; thus there is a supply of $CO_2$. Ocean temperature lagging atmosphere causes cloudier conditions and the extra $CO_2$ is absorbed, for a while, but eventually, as the ocean temperature is reduced, less $CO_2$ is absorbed and the atmosphere temperature rises above that of the ocean to produce a period of clear skies. The measured atmosphere-temperature profile and projected ocean temperature for the last major transition from Bell, (2003a) is shown here in figure 5. It may be observed that a temperature crossover apparently occurred, as expected to usually be the case, about 15,000yr after onset of glaciation.

For modest glaciation conditions, such as those before 800,000yr BP, this could have been what terminated glacial stages 15,000 to 20,000yr after the start of most glacial stages. What happens depends on the potential heat input. Before 800,000yr BP ice caps had not yet developed. Without nodal enhancement there was only modest capacity to sustain feedback cooling and glacial stages could be ended early by the mechanism just described. With both interglacial and glacial stages being limited in duration by the described mechanisms, reasonably uniform cycling lasted a long time, almost a million years in fact, before the advent of the grand node (figure3).

For recent cycles with ice caps in place and consequently greater feedback cooling potential, glaciation survived the initial fluctuating climate during glacial stages and grew to the point where something more drastic had to happen. As we have seen in the analysis above, since 400,000yr BP, a summer cycle coincident with summer in-sync was initially sufficient to end the glacial regime. It later became necessary to have the ocean start to freeze.

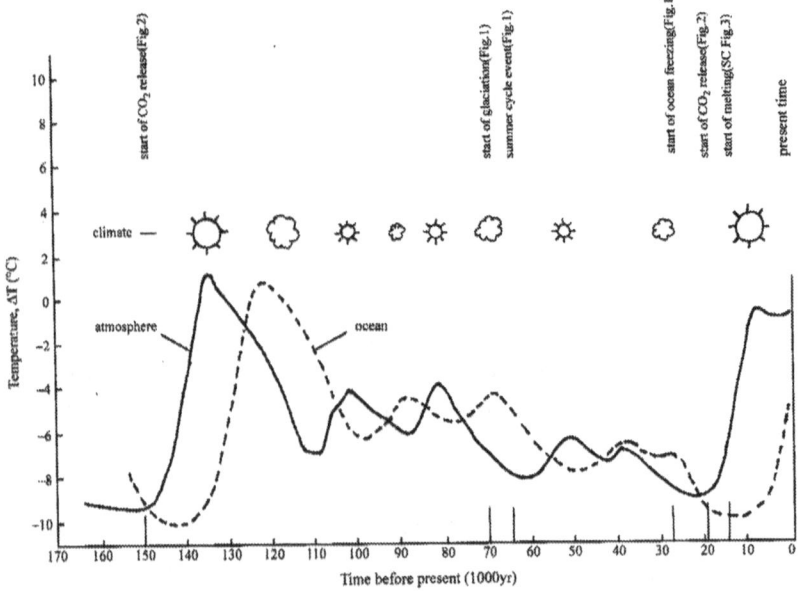

*Figure 5. Temperature profiles for the ice surface of Antarctica from Jouzel et al (1987) (modified to represent atmosphere) and that projected for the ocean for the last 160,000yr (from Bell (2003a). Expected climate conditions are shown as well as certain transitional features. Of particular interest is the period of relatively clear skies 15-20,000yr after the start of glaciation.*

A brief review of history will emphasize a point. It has been known since the time of Isaac Newton (1642-1727) that precession is a gyroscopic effect caused by gravitational forces of the Sun and Moon acting on the oblateness of the Earth's spherical shape (Ekman, 1993). James Bradley in 1748 was the first to observe that the orbit of the Moon causes a precession-related change in obliquity. Jean le Rond d'Alembert (1717-1783) was the first to attempt a calculation of the change in obliquity caused by the Moon, Leonhard Euler (1707-1783) made another attempt, including a term for the Sun's effect, and the problem was more or less solved in 1799 by Pierre de Laplace (1749-1827). Indications are from this, and from the calculations done by Milankovic, that precession should be the most important Earth-attitude cycle as far as climate change is concerned. The present work confirms that indeed precession is important. It engenders winter and summer climate cycles and all three interact with other cycles, namely orbit eccentricity, nutation and ocean temperature, to cause periods, well spaced out over time, of enhanced or reduced glaciation. The point is that we have been able to relate all anomalies to ocean temperature and $CO_2$ related effects or to the interaction of precession and other less effective cycles of constant period. No other cycles of greater influence are known to exist and therefore nothing further is required to explain the glaciation record.

A cursory search of the literature has not revealed any known connection between precession and orbit cycles. The very close match of in-sync effects, with glaciation evidence for 1.6 million years that has been found in the present work, suggests that the two indeed be related. Since gravity effects of the Sun/Earth/Moon system are shown to determine the period of precession and the precession-related change in obliquity perhaps they could also be shown to cause and maintain the orbit eccentricity cycle. All orbiting bodies in the Solar System have to be interrelated and it is therefore no surprise to find that the orbit and precession cycles are related. It is a surprise however to find that they are synchronized and exhibit harmonic beats.

The Earth's climate may be thought of as a vibrating system and, as is usual in a vibrating system, the frequency of component parts is modified slightly to maintain harmony. This is presumably why it was initially determined from harmonic nodes found with ocean sediment glaciation data, that the ocean temperature lags that of the atmosphere by 11,750yr (Bell, 2002), whereas at the beginning of an interglacial period the lag time is found to be only about 10,800yr. (Actually the lag time as determined in Bell, 2005a for recent major cycles using ice-core $CO_2$ and temperature data (Barnola et al, 1991) was 12,000yr. However times for ice core data are 11% higher than those obtained for ocean sediment data. Since the ocean sediment core timing is considered more accurate (Bell, 2002) this gives a corrected lag time of 10,800yr for recent large glaciation-induced cycles). At 11,750yr the ocean/atmosphere temperature cycle stays in tune with the winter precession and 41,000yr nutation cycles. With this line of thinking, it is also perhaps not a coincidence that the grand node thus produced, occurs at the same time as an in sync event involving orbit eccentricity and precession cycles. All components of the system are influencing the others to remain in harmony.

## 5 CONCLUSIONS

5.1 It is confirmed that the period of glacial/interglacial cycles is determined by a major ocean temperature cycle. Earth attitude cycles and other earth-bound effects have some influence over a longer term because they induce changes in the ocean cycle. As a result the duration of glacial/interglacial cycles is not constant and has increased from about 95,000 to 125,000yr during the last four cycles, the period only coincidentally averaging about the same as that of the orbit eccentricity cycle.

5.2 The idea that $CO_2$ is released and reabsorbed by the ocean in each glacial/interglacial cycle and that the $CO_2$ cycle is regulated by biomass intervention is supported by this analysis. The duration of interglacial stages is fairly constant over recorded history, their duration being determined, independent of other factors, by the time required for $CO_2$ to be released and reabsorbed. The duration of glacial stages, on the other hand, increases considerably as the rate of glaciation increases. The reason for this is that ocean temperature has to be induced to drop below that of the atmosphere, so that $CO_2$ release can begin, and this becomes more difficult as the rate and amount of glaciation increases.

5.3 Winter is now coincidentally going through precession minimum at perihelion of the orbit ellipse at maximum eccentricity and in this circumstance is said to be *in sync*. It is theorized that summer and winter alternate being in sync at 400,000yr intervals. Glaciation data confirms, though not dramatically, the theoretical timing of in-sync seasons. When summer is thought to have been in sync, glaciation is subdued and when winter is in sync, glaciation is enhanced. The main Earth-attitude cycles of precession and orbit produce glaciation nodes only at 800,000yr intervals. This finding adds further strength to arguments made elsewhere that the period of recent glacial/interglacial cycles is not directly related to the 100,000yr orbit-eccentricity cycle.

5.4  A confluence of several harmonic nodes of astronomical, and earthbound cycles, as well as an in-sync event occurred at 800,000yr BP. The confluence of so many effects seems to suggest that the Earth's climate operates as a vibrating system with various components adjusting themselves to remain in harmony with the others. It is probable that ice sheets were deposited on Antarctica and Greenland during the very cold period after 800,000yr BP.

5.5  No cycles of constant period, other than precession and its interaction with orbit eccentricity, nutation, and ocean temperature, are required to explain the observed anomalies in glaciation data over the last 1.6myr. It is fortuitous that northern season precession maxima cannot be in phase with the orbit ellipse perihelion at maximum eccentricity since then variation in the amount of glaciation would be much more spectacular. (It is possible for an in-phase effect in the Southern Hemisphere and since it is currently in phase the southern summer Sun should be hotter than the northern summer Sun).

Acknowledgements

The author gratefully acknowledges kind permission by *Quaternary Research* to republish data in figures 2 and 3. Advice and support from J. C. Anderson and editorial comment by J. A. L. Robertson, R. Bishop and A. Berger is also greatly appreciated.

References

Barnola JM, Pimienta P, Raynaud D, Korotkevich YS (1991) $CO_2$–Climate Relationship as Deduced from the Vostok Ice Core: a Re-examination Based on New Measurements and on a Re-evaluation of the Air Dating. Tellus 43B: 83-90.

Bell LG (2002). World Ocean Temperature Lag Time: An Analysis Based on Glaciation Data for the Last Two Million Years. Theor. Appl. Climatol 73 (3-4): 243-247.

Bell LG (2003a). Ice Age Mystery: A Proposed Theory for the Cause of Long Term Climate Change. Theor Appl Climatol 74 (3-4): 245-253.

Bell LG (2003b). A 30,000yr Precession-Related Cycle Affecting Climate. Theor Appl Climatol 74 (3-4): 255-271.

Bell LG (2005a). Ocean/Atmosphere Temperature Disparity Effects. (in preparation).

Bell LG (2005b). Interpretation of $CO_2$ Data with the Ocean as Source and Sink. (in preparation).

Berger A (1977). Long-term Variation of the Earth's Orbital Elements, Celestial Mechanics 15: 53-74.

Berger A and MF Loutre (1991). Insolation Values for the Climate of the Last 10 Million Years. Quaternary Science Reviews !0: 297-317.

Bishop R. (2001). Some Astronomical and Physical Data, in The Observers Handbook, edited by R. Gupta, Royal Astronomical Society of Canada, Toronto University Press, Toronto: 25.

Ekman MA. (1993). Concise History of the Theories of Tides, Precession-Nutation and Polar Motion (from Antiquity to 1950). Surveys in Geophysics: Dordrecht, The Netherlands, 146(6): 585-617,

Jouzel J, Lorius C, Petit JR, Grenthon C, Barkov NI, Katliokov VM, Petrov VM (1987) Vostok Ice Core: a Continuous Isotopic Temperature Record over the Last Climatic Cycle (160,000 years). Nature 329: 403-40

Porter SC (1989). Some Geological Implications of Average Quaternary Glacial Conditions. Quaternary Research 32: 245-261, Figure 1: b and c.

This latest study provides still more evidence for, and confidence in, the new theory of climate change. Overwhelming evidence in this authors view. To explain the glaciation record it is found necessary to reintroduce the ocean temperature cycle and its effect on release and absorption of greenhouse $CO_2$ as well as the concept of biomass intervention. Unfortunately this does not bode well for the future.

# Chapter 8
## Future Implications

It is very dangerous to be releasing $CO_2$ into the atmosphere. In a nutshell, according to the developing theory, it heats the upper atmosphere and causes skies to become clearer. Thus not only is the planet heated by a greenhouse effect but also by less cloud to reflect heat back into space. Clearer skies, though generally producing higher temperatures, also unfortunately have the effect of reducing atmosphere temperatures in the far north during winter, because fewer clouds allow more heat to radiate into space. This causes $CO_2$ to diffuse out of the northern oceans and to thus maintain the excess buildup in the atmosphere that comes from human activity. In the long run, without significant potential biomass to intervene, a feedback mechanism will cause a natural buildup to continue after the industrial age has ended. This is an effect of the current continental configuration and there is nothing we can do about it.

Also, according to the theory, the climate will continue to deteriorate as $CO_2$ increases. Warmer air for most of the planet interacting with colder air generated in the north will cause more violent weather. Evaporation from warmer surface water will provide more energy for the violent interaction. As I write in December 2006 the coast of British Columbia is suffering the third of three severe windstorms in the last month and in Tasmania and mainland Australia wildfires are raging; a taste of what we are in for.

The report inserted below includes an attempt to estimate what will happen long-term. While considering the implications, previous reports were studied again and several misconceptions and errors were discovered. Thus, along with predicting the future, some effort is made to correct these faults in this final report.

# 8. Indicated Implications of Global Warming Based on a Theory of Climate Change with the Ocean as Source and Sink for $CO_2$
## L.G. Bell

*Abstract*

*A new theory on the cause of climate change has been developed using $CO_2$ and temperature data for both past and present and glaciation data going back 1.8myr. The theory is used in the present report to speculate on future developments. Temperature and $CO_2$ records are reexamined and an attempt is made to extrapolate. Data suggests that the Earth's temperature is unaffected by glaciation however this could be due to the fact that measurements are made on ice cores from Antarctica. It is also observed that only 55% of anthropogenically produced $CO_2$ remains in the atmosphere and that world temperature seems to lag buildup by at least 15yr. These new findings are not critical to the outcome however it is now realized that natural release rate in past glaciation cycling was not kept to a minimum by lack of supply, as previously assumed, but rather by eventual heating of the ocean. Even though a modest view is taken the implications of this latest finding are that post-industrial age $CO_2$ levels will increase exponentially in the future and that the immediate rate of rise will increase as the starting level increases. It is also predicted that as $CO_2$ levels increase the climate will continue to deteriorate.*

## 1.Introduction

During recent major glacial/interglacial cycles the $CO_2$ concentration in the atmosphere has cycled by as much as 95ppmv. This amount, along with the resulting clearer skies, has been enough, several times in the past, to make the change from Ice Age conditions to those of balmy interglacial periods. Recent anthropogenic increase from the burning of fossil fuels has caused an increase in atmospheric $CO_2$ concentration of about the same order of magnitude; from about 275ppmv before the start of the industrial revolution to current readings of 375ppmv at Mauna Loa observatory (Conway et al, 2005). This has to be of some concern. Should we expect imminent change to be the same order of magnitude as that of glaciation cycles? Should we allow the apparent modest changes, the result of global warming, to assure us that nothing much worse will happen, or should we regard them as a harbinger of drastic change to come?

A new theory is being developed for the cause of climate change. The theory has been based on analysis of Antarctic ice-core and ocean sediment data as well as remote site observations. An important tenet of the new theory is that the ocean acts as source and sink for $CO_2$ and change takes place as a result of thermal diffusion. The purpose of the current exercise is to use the theory to predict what climate conditions might eventually become evident if we should stop fossil fuel consumption at the current $CO_2$ level and also if we should not stop now but do so at higher levels. As is usual there are other things to discover along the way.

## 2. Information

2.1 The relationship between atmospheric $CO_2$ concentration and Antarctic temperature is well documented from ice core measurements (Barnola et al, 1994). Temperature data over time was obtained using hydrogen isotopes and show that the temperature change on the Antarctic surface at the start of an interglacial period, the result of a 95ppmv increase, is about 0.95°C, the ratio between the two being thus about 100 to 1 in corresponding units. (Note: The temperature change as measured on the Antarctic surface has been taken in this investigation to be as-stated, about 1° C for 100ppmv increase in carbon dioxide. The decimal points are indistinct on the original data of Barnola et al and close examination reveals that we are out by a factor of 10. 100ppmv increase causes a 10°C increase in ice-core readings on the Antarctic surface according to the published data. We continue however to use the lower figure assuming a factor of 10 difference between poles and the world as a whole).

2.2 It is reported by the World Meteorological Organization (Neilsen et al, 2002) that average global temperature has increased as a result of greenhouse warming by 0.8°C. Anthropogenic increase in $CO_2$ concentration was about 93ppmv in 2002 making it possible, but not certain, that there is again an approximate 100 to 1 ratio.

2.3 It is theorized in other reports in this series (e.g. Bell, 2005a) that increasing $CO_2$ in the atmosphere causes skies to become clearer. As a result of this, during the early part of the previous interglacial period, Antarctic surface temperature increased in proportion to atmospheric

## Implications

$CO_2$ concentration, with about four parts being attributable to greenhouse warming and one part to loss of albedo, the result of clearing skies. (It is now noticed that there seems to be no change in albedo associated with melting ice, leading one to think that variation in cloud cover is much more important. Data is collected using ice cores from Antarctica. It could be that the albedo at this location is determined only by cloud cover, unaffected by the amount of glaciation mainly in the northern hemisphere). It is speculated in other work (Bell, 2005c), that increasing ice had to be offset by more $CO_2$ in the atmosphere to achieve climate control in the far distant past. What we are observing during the interglacial heat-up stage is that more $CO_2$ causes less ice, and less cloud, but of course during this time the climate is not under control so the temperature keeps rising.

2.4 According to the developing theory, buildup in the atmosphere at the start of the interglacial period is caused by release of $CO_2$ from northern oceans by means of a diffusion mechanism. Even though global temperature increases with clearing skies, the result of increasing $CO_2$, northern winters get colder thus increasing the local release from the ocean. This would not have happened in the distant past, but it does now because of the current continental configuration. The buildup with time is initially a function of $t^4$, reaching a rate of about 12.5ppmv/1000yr at about 40ppmv additional $CO_2$. The buildup then slows to a linear rate of 10ppmv per 1,000yr (Bell, 2005a). It has been speculated in previous reports in this series that upper layers of the northern oceans, where release is occurring, become depleted by rapid release before 40ppmv and subsequent release depends on a linear rate of transport of $CO_2$ from the ocean depths.

2.5 It is found (Bell, 2005a) that ocean/atmosphere temperature disparity may be used to calculate the natural release, or absorption, of $CO_2$ by the ocean during glaciation cycles. Using simple integration the formula for the contribution of a 2000yr period is 22.6ppmv for each 1°C disparity. Disparity for use in these calculations is determined from Antarctic ice core data. Atmosphere temperature is as measured and ocean temperature is projected from atmosphere temperature curves assuming a 12,000yr lag (See Bell 2002 where ocean lag time was determined to be 11,500yr).

149

2.6 The buildup of $CO_2$ in a normal interglacial period is stopped at around 90ppmv by the intervention of the biomass. It is believed that, as $CO_2$ gradually increases, forests re-inhabit the bare rocks and gravel of northern plains and mountains. Once a certain level of $CO_2$ is achieved, the concentration is held more or less constant by biomass growth, and eventual decay, for 20,000yr or more (Bell, 2003a). This hiatus allows the ocean temperature to "catch up", which sets the stage for subsequent major change.

2.7 Before the start of the industrial revolution the $CO_2$ concentration was 275ppmv, then and still today, the ocean is expected to be about 0.4°C cooler than the atmosphere (Bell, 2003a). This is arrived at by the idea that the ocean is lagging behind the temperature increase at the beginning of the interglacial period.

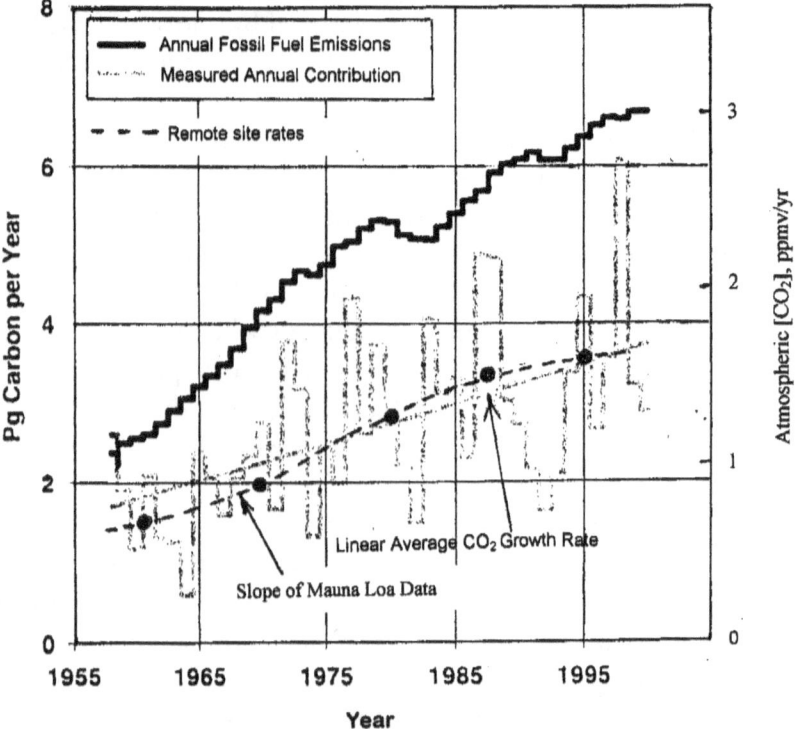

*Figure 1. Estimated annual fossil fuel emissions of $CO_2$ and the globally measured annual contribution to atmospheric $CO_2$ concentration since 1957 (Conway et al, 2003). The Linear Average $CO_2$ Growth Rate and data it is based on, is suppressed to background. Also shown is the slope*

# Implications

*of growth curves obtained at Mauna Loa and Antarctica (from Keeling, 2001).*

2.8 It is indicated that only about half of fossil fuel emissions of $CO_2$ stay in the atmosphere, the other half presumably being absorbed into surface layers of the ocean to balance that in the atmosphere (background in figure 1 from Conway et al, 2003).

2.9 Atmospheric $CO_2$ concentration measurements have been made by CMDL at the Mauna Loa and Antarctic Observatories. The famous Keeling curve is published in various journals and encyclopedia (e.g. Keeling CD, 2001). Extension of the Mauna Loa atmosphere $CO_2$ measurements to the present also indicates that the average concentration currently stands at about 375ppmv (Conway ET al, 2005).

3. Analysis

3.1 Temperature

There appears to be agreement, in item 2 above, between the past Antarctic surface and estimated global conditions, each increasing 1°C for a 100ppmv increase in $[CO_2]$. There is the problem, mentioned above, that Antarctic ice core data shows an increase 10 times as much. It is also fairly certain that the global increase of 0.8°C estimated by Neilsen et al is not an equilibrium amount for the induced change. It should take time to settle out. We take a chance assuming a 1 to 100 relationship but it is the best we can do and it is an optimistic view, as we will see.

We are accustomed to having conditions outside the window change from season to season, from day to day and even from hour to hour. We are led by this to think that average global temperature would respond almost instantaneously to changing conditions but this may not be so. Taking an extreme example, it was found that very slight changes in heat input, the result of precession and other Earth-attitude and Earthbound cycles, changes the amount of glaciation over periods of more than 4 million years (Bell, 2002). A feedback cooling effect of glacial ice is perhaps important in this scenario. However it is implied by this extreme case that there is such a thing as long-term heat retention, by the ocean and by the land perhaps to depths of hundreds of meters and even to the very core of the Earth. Another relevant observation (Bell, 2003b) is that slight additional heat input from precession-related, 30,000yr "summer" cycles acting over many thousands of years can change the rate of glaciation and even bring a glacial stage to an end (Bell, 2005d). It is also found necessary, in order to explain temperature and $CO_2$ profiles of recent glaciation cycles, to have the ocean temperature lag that of the rest of the environment by about 12,000yr. Should the rest of the environment not also lag?

Surface temperatures vary widely diurnally and annually but changing the global average should be expected to take longer. Modest approach to equilibrium apparently does happen much more quickly than the long-term glaciation and other effects mentioned above would tend to suggest. During the heat-up period at the start of an interglacial period, Antarctic surface temperature seems to follow $CO_2$ buildup fairly well. However the introduction of 95ppmv takes place over about 16,000yr, whereas we have introduced the same amount in only 300yr. To make an analogy, one would not expect to immediately achieve an equilibrium temperature upon turning the element under a pot of water to a low setting. It takes time for a balance between heat input on one hand and heat lost by convection, black body radiation and evaporation on the other. Something even more complicated happens with changing $CO_2$ in the atmosphere. With global warming there is the greenhouse effect from increasing $CO_2$, reduction of albedo from clearing skies, melting glaciers, forest clearance, etc. and diffusion of the extra heat into land, sea and ice. There are also some areas that actually become colder, at night and especially in high latitude winter, because loss of cloud cover allows heat to radiate into space. No doubt global temperature does lag $CO_2$ buildup, perhaps by hundreds of years.

There is actually evidence that the lag time, or at least the greater part of it, may be discernable and of relatively short duration. Advantage is taken of the marked increase in the rate of $CO_2$ buildup that occurred at about the year 1960 (Berner et al, 2001). Global temperature data shows a great deal of scatter (Neilsen et al, 2003), however taking some license with that and liberal curve fitting it appears that temperature may lag $CO_2$ buildup by only 15yr or so, as shown in Figure 2. According to Einstein's formula for vertical transport (Jacob, 1999) it takes 5-10yr for atmosphere molecules to be transported to the stratosphere. $CO_2$ molecules being larger would presumably take longer thus perhaps a 15yr delay of global warming should be expected for this reason alone. Incidentally the hiatus in $CO_2$ buildup in the 1940s and 50s, before atmosphere measurements began, would be the result of thermal diffusion into the ice on Antarctica during southern summer (see Bell 2005b).

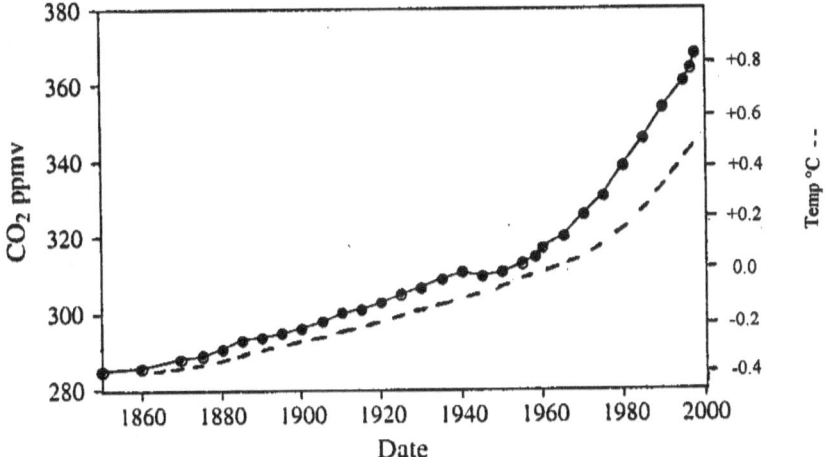

*Figure 2. Atmospheric $CO_2$ concentration and world temperature since 1850 showing approximate temperature lag time of 15yr.*

## 3.2 $CO_2$ concentration

### 3.2.1 Biomass content

It was estimated in a companion report (Bell, 2005a) that the total biomass of the planet is about 670ppmv, should it be introduced to the atmosphere. This figure for total biomass was based on the maximum amount, 67ppmv, retained in the biomass during the previous interglacial period. It was calculated using the disparity release formula and taken to be 1/10 of the total biomass of the planet.

It is realized that the 67ppmv figure for displaced northern biomass should perhaps be doubled to 134ppmv. During the normal buildup of an interglacial period the disparity formula estimates only the amount of $CO_2$ released to the atmosphere and neglects that being reabsorbed by the ocean. Once the biomass intervenes the calculated amount should be augmented above that obtained by the formula because the amount being absorbed by the biomass does not have to be balanced by an amount in the ocean surface layers. As a consequence of this thought, perhaps a better figure for total biomass, again assuming 10% for northern biomass, would be 10x134 = 1340ppmv. (4000PgC or 1,800ppmv is the value generally accepted by others (Jacob, 1999)).

As for the current and future buildup one has to consider the contribution from destruction of the biomass. It is noted that the increase being measured at Mauna Loa includes emissions from all sources, whether it be burning fossil fuel, degradation of the biomass, diffusion from the ocean, or something else.

It is mentioned above that approximately half of fossil fuel emissions of $CO_2$ are retained in the atmosphere. This was based on the *linear average $CO_2$ growth rate* of global atmospheric data shown in the background of figure 1 (CMDL, 2003). It is now realized that it is possible to get a more accurate fraction. Rate of increase may be determined by measuring the slope of $CO_2$ buildup at Mauna Loa and Antarctic sites. The result of this procedure is shown as a curve with data points in figure 1. The percentage retained in the atmosphere is easily calculated and remains fairly constant over time with the average amount retained being about 55%. Since the ratio remains constant, either the biomass contribution is small, or it increases in proportion to the fossil fuel contribution. If it were true that half of anthropogenically generated $CO_2$ stays in the atmosphere, and that the other half is taken up by the ocean to balance, the extra 5% would be half the amount that comes from destruction of the environment.

### 3.2.2 Predicted future increase

Let us assume that we stop burning fossil fuel with the concentration of 375ppmv presently in the atmosphere.

Because of the temperature disparity naturally present and that introduced by adding $CO_2$, the concentration would unfortunately continue to increase, at a lower rate to be sure, but nonetheless a rate that is significant and unrelenting. According to recent work (Bell, 2005a) the maximum rate we should expect is 10ppmv per 1000yr, since this is what was observed for the major part of the buildup at the beginning of the previous interglacial period. It is now realized however that it is temperature disparity that is important and that the linear buildup being constant was the result of the ocean heating rather than lack of supply as speculated in Bell, 2005a. Although the ocean temperature is lagging, by the time the constant rate was in effect it was heating at the same rate as the atmosphere, disparity therefore remained constant, and the rate of release stayed constant.

This latest discovery has significant implications for the future because buildup will not be inhibited in any way. The biomass cannot intervene, in fact may add to the buildup, and now we find that the rate of release cannot be controlled and is subject to a feedback mechanism. Temperature disparity causes release of $CO_2$ and this increases temperature disparity and so on.

It should be possible, by using disparity $\delta$ as described in section 2.5, to calculate $CO_2$ release. As noted above, 100ppmv causes atmosphere temperature to rise about 1°C. Since disparity rules, we may predict the future by assigning 1° disparity for each addition of 100ppmv $CO_2$ above the pre industrial 275ppmv and get a value for the ocean from figure 5 of Bell, 2005d. Disparity at 375ppmv for example is currently about 1.4°C (0.4 for the ocean and 1.0 for the addition of 100ppmv). By the formula:

$$\Delta \, [CO_2] \text{ increase for 2000yr} = 22.6 \times 1.4 = 31.6\text{ppmv}$$

By the end of the 2000yr period the concentration will thus have increased to 375 + 32 = 407ppmv. For the second 2000yr period disparity is 1.32 for the atmosphere plus 0.2 for the ocean = 1.52, which multiplied by 22.6 gives a contribution for this time of 34.4 to make a total of 476ppmv. By the third 2000yr period ocean disparity is gone and further increase is by simple exponential growth. $\delta$ for the third period is 476-275/100 = 2.01 x 22.6 = 45.4 and [$CO_2$] becomes 521ppmv.

Figure 3 shows the result of calculations done in this way for starting concentrations of 375ppmv, 475ppmv and 1000ppmv up to 12,000yr into the future but with the ocean contribution only 0.2 for the first 2000yr period and 0.0 for the second. Comparison with the calculations in the text above emphasizes the effect, of even what might be considered insignificant difference, on the projected buildup. Exponential increase with time becomes more dramatic as starting level increases. Increasing rate depends on skies becoming clearer, thus a possible mitigating factor is that skies can only become so clear. Rate of change would then be linear. This is unlikely and it is also unlikely that eventual ocean heating could significantly limit release.

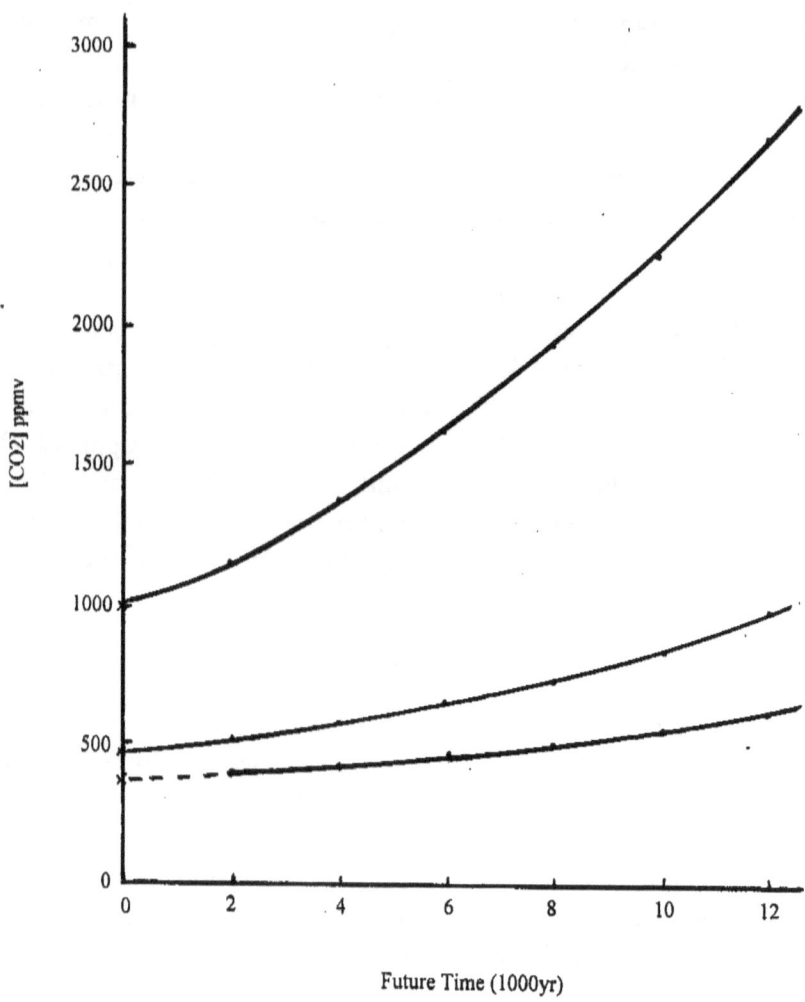

*Figure 3. Expected spontaneous change in post-industrial age atmospheric $CO_2$ concentration with starting levels of 375, 425 and 1000ppmv. Calculations were done using the disparity formula for release with $1°C$ disparity assigned to each 100ppmv above 275ppmv.*

### 3.2.3 Sources and Limits

Having gone through this exercise it is realized that the continuing extra contribution has to come from somewhere. One cannot argue that it comes from the surface layers of the ocean because the concentration there has to increase along with that in the atmosphere to provide a balance across the interface. Release is restricted to a certain area of the northern oceans and, in that area levels in the ocean also have to increase by almost the same amount as the rest of the ocean. Replenishment must come from below during winter months as shown in figure 4. Some fraction of whatever amount is released from northern oceans will be retained in the atmosphere while some is reabsorbed, by all surface areas of the ocean, to maintain a balance across the interface.

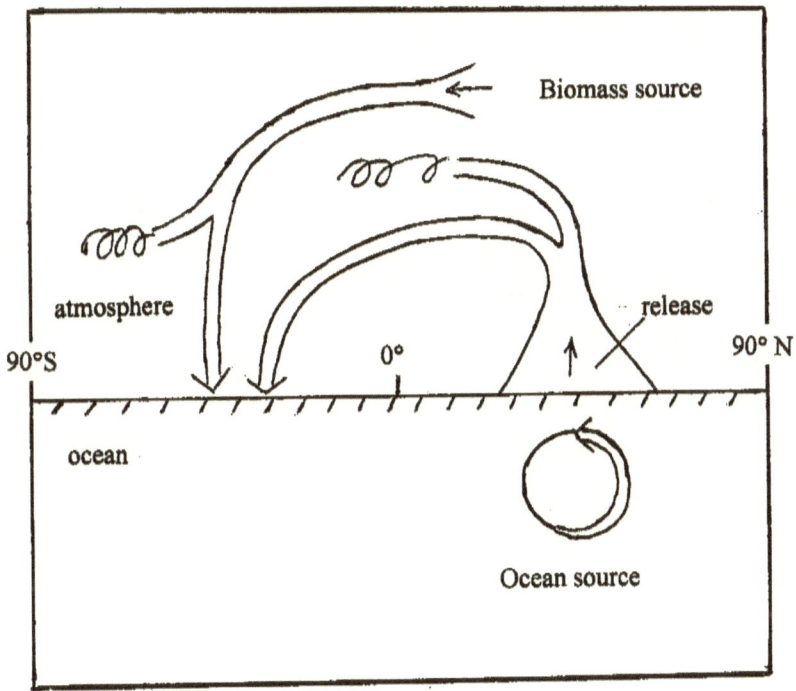

*Figure 4. Stylized depiction of proposed mechanisms for increasing $CO_2$ in the post-industrial atmosphere. About 55% of that from both diffusion release and biomass destruction stays in the atmosphere.*

The biomass source is not locked in permanently and in the future, as temperature and $CO_2$ concentration increase, there will be more photosynthesis but also more wild fires; one reducing global warming, the other increasing it. The thought is that the wild fires will win and provide an extra contribution to $CO_2$ growth (included in figure 4).

Since there is no prospect of biomass intervention, as occurs normally in glaciation cycles, release should continue, eventually achieving a high equilibrium atmospheric concentration. Berner et al

(2001) set this figure at 5,000ppmv for a similar scenario 550myr BP.
Berner's estimates were made with the premise that biomass acts as the
main source and sink for $CO_2$, but no mechanism is described for the
buildup, or for a regulating mechanism that would provide a limit. It is
theorized in current work that, at the level of 1000ppmv, if approached
gradually, humidity should increase sufficiently so as to cause clouds
to form in the stratosphere and thereby stop further increase (Bell,
2005c)). If we should increase the starting level to 1000ppmv it could
stop there or continue to increase exponentially with time as shown in
figure 3.

## 3.3 Weather

It is perhaps even greater folly to attempt to predict what the
future weather will be. It is speculated in this series of reports that
as $CO_2$ concentration increases, skies get clearer. At the start of an
interglacial period it is found, for example, that $CO_2$ is released from
the northern Pacific Ocean at an increasing rate with time (Bell, 2005a).
The explanation for this is that, as skies get clearer winters get colder
so release rate increases. Meanwhile summers are getting warmer. The
interaction of cold and warm air is what determines weather conditions,
the more the difference increases the more violent the weather becomes.
The greatest change in the future will come as a result of the assault
imposed by continued fossil fuel consumption. Once the industrial age
ends, conditions will continue to change, but at a much-reduced rate. As
$CO_2$ increases, whatever the route followed, the climate will continue to
become more violent and unpredictable. As well as increasing disparity
there will also be more evaporation as the ocean surface heats up.
Increasingly there will be more incidence of excessive precipitation, more
rain in some areas, more flooding, more windstorms, more snowstorms,
more ice storms, more tornadoes, more hurricanes etc. There will also
be more episodes of severe drought, both localized and widespread.

Wild fires will also break out more often as $CO_2$ increases and
especially along the western shores of North America and Europe. The
reason for expecting this local effect is that $CO_2$ is released by northern
oceans in late winter and early spring when forests are most vulnerable.
Light on-shore winds can occasionally increase greenhouse warming
and hence the incidence of fire inland along the coast. The same area in
particular and indeed most mid latitudes will also experience increasing

episodes of torrential rain. In spite of a moderating influence of the southern ocean, fires will also increase in frequency in Australia especially in the eastern states, because the Sun should be somewhat hotter in the Southern Hemisphere (Bell, 2005d).

4. Conclusions

4.1 It is now observed, from data for the previous interglacial period, that albedo on Antarctica is unaffected by glaciation elsewhere.

4.2 Global temperature increases in proportion to $CO_2$ concentration but lags by an unspecified time, maybe 15yr, but more likely many hundreds of years.

4.3 A rate measurement for $CO_2$ increase shows that 55% of that released by burning fossil fuels is retained in the atmosphere.

4.4 In the eventual post-industrial age, $CO_2$ concentration in the atmosphere should continue to increase exponentially with time and, as the anthropogenically induced levels increase, future buildup will be more dramatic. Increasing stratospheric cloud cover may limit release but we should not expect biomass intervention to come into play as it has in the normal glaciation cycling of the past few million years.

4.5 Climate is linked to $CO_2$ concentration and weather conditions should therefore become increasingly more violent and unpredictable as it rises.

References:

Barnola JM, D. Raynaud, C.Lorius and Y. S. Korotkevich, 1994. Historical Record from the Vostok Ice Core. *Trends '93A*: 7-10.

Bennett J (2003). Presentation to Brampton Chapter of Professional Engineers of Ontario. Also Environment Canada, An Introduction to Climate Change, A Canadian Perspective ISBN #: 0-662-41247-8, October 2005, Figure 2.4

Bell LG. 2002. World Ocean Temperature Lag Time: An Analysis Based on Glaciation Data for the Last Two Million Years. *Theoretical and Applied Climatology*. 73 (3-4): 243-247.

Bell LG (2003a) Ice Age mystery: a proposed theory for the cause of long term climate change. Theor Appl Climatol 74 (3-4): 235-244.

Bell LG (2003b). A 30,000yr Precession-Related Cycle Affecting Climate. Theor Appl Climatol 74 (3-4): 255-271.

Bell LG. 2005a. Ocean/Atmosphere Temperature Disparity Effects: An Analysis of $CO_2$ and Temperature Data for the Previous Interglacial Period. (In preparation).

Bell LG. 2005b. Interpretation of $CO_2$ data with ocean as source and sink. (In preparation).

Bell LG. 2005c. Earth History with the Ocean as Source and Sink for $CO_2$. (In preparation).

Bell LG. 2005d. Evidence of an Effect of the Earth Orbit Cycle on Glaciation. (In preparation).

Berner, E.K., and Berner, R.A. 2001.Carbon Dioxide Concentration and Climate over Geological Times. Munn R.E., (Ed.), *Encyclopedia of Global Environmental Change,* John Wiley and Sons, Chichester, 1: 249-254.

Conway T. and Tans P. 2003. *Climate Monitoring and Diagnostics Laboratory,* NOAA http://www.cmdl.noaa.gov/ publications/ annrpt26/index.html.

Conway T. and Tans P. 2005. *Climate Monitoring and Diagnostics Laboratory,*NOAA ftp://ftp.cmdl.noaa.gov/ccg/co2/in-situ/

Jacob DJ (1999) *Introduction to Atmospheric Chemistry.* Princeton University Press: Princeton.

Keeling CD. 2001. Keeling, Charles David. *Encyclopedia of Global Environmental Change,* Munn RE ed. 1: 484-485. John Wiley and Sons: Chichester.

Nielsen E and Polli E (2002) World Meteorological Organization Graphic, as published in the Toronto Star, Jan.3, 2003

## Implications

I heard, soon after undertaking this project, that someone had proposed a plan to cover the northern Pacific Ocean with Ping Pong balls in an effort to offset the problem of greenhouse warming. Of course I thought at the time it was a joke and immediately rejected the idea. Now it makes me think that certain other people, and perhaps climatologists everywhere, believe, as I have been maintaining, that that is where the problem is and that it may not be solvable. That is perhaps why journals do not tolerate analysis. They would be accepting the inevitable conclusion. No one wants to do this and I hereby forgive all those reviewers for the nasty comments and putdowns. I don't want to accept it myself and send this book to print with trepidation. I take comfort and justification from the old adage - If you don't acknowledge a problem you can't hope to fix it. Please be aware that I freely admit to not being an expert in climatology, or any other field of scientific endeavor, and would be very pleased if someone could show that I am completely wrong. GB

------- X X -------